Artificial Intelligence and Spectroscopic Techniques for Gemology Applications

Online at: https://doi.org/10.1088/978-0-7503-3927-8

IOP Series in Spectroscopic Methods and Applications

Series Editor
Ashutosh Kumar Shukla, Ewing Christian College, Prayagraj, India

About the Series

The *IOP Series in Spectroscopic Methods and Applications* aims to develop a library of key texts encompassing the broad range of spectroscopic techniques, instrumentation and fields of applications found in modern science. Some examples of the spectroscopic techniques incorporated include, IR, UV/VIS, Raman, EPR, NMR, fluorescence spectroscopy, force spectroscopy, hyperspectral imaging, neutron scattering, Mössbauer spectroscopy, photoemission spectroscopy and all the various specialist techniques within these broad domains and more. Microscopic characterization techniques may also be included where they are used in conjunction with spectroscopic methods for sample analysis.

The series contains two broad types of approach. Those addressing a particular field of application and reviews the numerous relevant spectroscopic methods applicable to the field, and those that focus on a specific spectroscopic method which will permit a greater in-depth review of the theory and instrumentation technology.

A full list of titles published in this series can be found here: https://iopscience.iop.org/bookListInfo/iop-series-in-spectroscopic-methods-and-applications.

Artificial Intelligence and Spectroscopic Techniques for Gemology Applications

Edited by

Ashutosh Kumar Shukla
Department of Physics, Ewing Christian College, Prayagraj, India

IOP Publishing, Bristol, UK

ISBN 978-0-7503-3927-8 (ebook)
ISBN 978-0-7503-3925-4 (print)
ISBN 978-0-7503-3928-5 (myPrint)
ISBN 978-0-7503-3926-1 (mobi)

DOI 10.1088/978-0-7503-3927-8

Version: 20221201

IOP ebooks

British Library Cataloguing-in-Publication Data: A catalogue record for this book is available from the British Library.

Published by IOP Publishing, wholly owned by The Institute of Physics, London

IOP Publishing, No.2 The Distillery, Glassfields, Avon Street, Bristol, BS2 0GR, UK

US Office: IOP Publishing, Inc., 190 North Independence Mall West, Suite 601, Philadelphia, PA 19106, USA

To my niece Anvesha.

Contents

Preface

This collection highlights the applications of different spectroscopic techniques to gemology. There are five chapters in this volume covering laser-induced breakdown spectroscopy, Raman spectroscopy, Fourier-transform infrared spectroscopy, x-ray fluorescence analysis and optical tomography.

Chapter 1, Laser-induced breakdown spectroscopy for gemological testing by Francesco Poggialini, Beatrice Campanella, Stefano Legnaioli, Simona Raneri and Vincenzo Palleschi, covers an overview of laser-induced breakdown spectroscopy and highlights the scope of this technique for the analysis of precious stones and gems minimizing surface damage to the samples. Chapter 2, Raman spectroscopy for the non-destructive analysis of gemstones by Danilo Bersani, Laura Fornasini, Peter Vandenabeele and Anastasia Rousaki, begins with a brief introduction to the technique and provides a compact overview of theory and applications with gemstone examples and effectiveness of mobile Raman systems. Chapter 3, Application of Fourier-transform infrared spectroscopy and machine learning algorithm for gem identification by Pimthong Thongnopkun, Kanet Wongravee, Prompong Pienpinijtham and Aumaparn Phlayrahan, begins with concepts of infrared spectroscopy and sampling techniques and covers gemstone analysis applications of the technique in detail. This chapter uniquely highlights the potential applications of Fourier-transform infrared spectroscopy coupled with machine learning. Chapter 4, A ruby stone grading inspection using an optical tomography system by Syarfa Najihah Raisin, Juliza Jamaludin and Fatinah Mohd Rahalim, talks about the analysis of transparency of ruby stones using image reconstruction. Chapter 5, Trace elements and big data application to gemology by x-ray fluorescence by Yujie Gao, Moqing Lin and Xueying Sun, mentions x-ray fluorescence applications to identify species and variety of gemstones including their geographic origin.

I sincerely thank Dr John Navas, Senior Commissioning Manager, IOP Publishing for giving me an opportunity to present this book to readers. I also wish to thank the IOP Publishing team for all the support during the development of this project. I thank the expert authors for enriching this volume with their academic contributions.

Ashutosh Kumar Shukla
Prayagraj, India
August 2022

Editor biography

Ashutosh Kumar Shukla

Ashutosh Kumar Shukla has more than two decades of physics teaching and research experience. He has had numerous articles and review articles published in peer-reviewed journals. He has also authored textbooks and participated as an editor in more than 20 edited volumes published by reputed publishers, often prepared in collaboration with experts from different countries. He intends to continue pursuing his academic interests, which include spectroscopy applications in different fields of societal importance.

List of contributors

Danilo Bersani
Department of Mathematical, Physical and Computer Sciences, University of Parma, Parco Area delle Scienze 7/A, 43124 Parma, Italy

Beatrice Campanella
Applied and Laser Spectroscopy Laboratory, ICCOM-CNR, Research area of CNR, Via G. Moruzzi 1, 56124 Pisa, Italy

Laura Fornasini
Institute of Chemistry of Organometallic Compounds, National Research Council (ICCOM-CNR), Via G. Moruzzi 1, 56124 Pisa, Italy

Yujie Gao
Guild Gem Laboratories, Shenzhen, Guangdong Province, China

Juliza Jamaludin
Department of Electrical and Electronic Engineering, Faculty of Engineering and Built Environment, Universiti Sains Islam Malaysia, 71800 Nilai, Negeri Sembilan, Malaysia

Stefano Legnaioli
Applied and Laser Spectroscopy Laboratory, ICCOM-CNR, Research area of CNR, Via G. Moruzzi 1, 56124 Pisa, Italy

Xu Li
The Testing and Technology Center for Industrial Product of Shenzhen Customs, Shenzhen City, Guangdong Province, China

Moqing Lin
Guild Gem Laboratories, Shenzhen, Guangdong Province, China

Fatinah Mohd Rahalim
Department of Electrical and Electronic Engineering, Faculty of Engineering and Built Environment, Universiti Sains Islam Malaysia, 71800 Nilai, Negeri Sembilan, Malaysia

Vincenzo Palleschi
Applied and Laser Spectroscopy Laboratory, ICCOM-CNR, Research area of CNR, Via G. Moruzzi 1, 56124 Pisa, Italy

Aumaparn Phlayrahan
Gems and Jewelry Research Unit, Faculty of Gems, Burapha University Chanthaburi Campus, 57 Chonlaprathan Road, T. Kamong A. Thamai, Chanthaburi 22170, Thailand

Prompong Pienpinijtham
Sensor Research Unit, Department of Chemistry, Faculty of Science, Chulalongkorn University, Pathumwan, Bangkok 1330, Thailand
and
National Nanotechnology Center of Advanced Structural and Functional Materials, Faculty of Science, Chulalongkorn University, Pathumwan, Bangkok 1330, Thailand
and
Center of Excellence in Bioactive Resources for Innovative Clinical Applications, Chulalongkorn University, Pathumwan, Bangkok 1330, Thailand

Francesco Poggialini
Applied and Laser Spectroscopy Laboratory, ICCOM-CNR, Research area of CNR, Via G. Moruzzi 1, 56124 Pisa, Italy

Syarfa Najihah Raisin
Department of Electrical and Electronic Engineering, Faculty of Engineering and Built Environment, Universiti Sains Islam Malaysia, 71800 Nilai, Negeri Sembilan, Malaysia

Simona Raneri
Applied and Laser Spectroscopy Laboratory, ICCOM-CNR, Research area of CNR, Via G. Moruzzi 1, 56124 Pisa, Italy

Anastasia Rousaki
Raman Spectroscopy Research Group, Department of Chemistry, Ghent University, Krijgslaan 281 (S-12), B-9000 Ghent, Belgium

Xueying Sun
Guild Gem Laboratories, Shenzhen, Guangdong Province, China

Pimthong Thongnopkun
Gems and Jewelry Research Unit, Faculty of Science, Burapha University, Saen Suk, Chonburi District, Chonburi 20131, Thailand

Peter Vandenabeele
Raman Spectroscopy Research Group, Department of Chemistry, Ghent University, Krijgslaan 281 (S-12), B-9000 Ghent, Belgium
and
Archaeometry Research Group, Department of Archaeology, Ghent University, Sint-Pietersnieuwstraat 35, B-9000 Ghent, Belgium

Kanet Wongravee
Sensor Research Unit, Department of Chemistry, Faculty of Science, Chulalongkorn University, Pathumwan, Bangkok 1330, Thailand
and
National Nanotechnology Center of Advanced Structural and Functional Materials, Faculty of Science, Chulalongkorn University, Pathumwan, Bangkok 1330, Thailand

IOP Publishing

Artificial Intelligence and Spectroscopic Techniques for Gemology Applications

Ashutosh Kumar Shukla

Chapter 1

Laser-induced breakdown spectroscopy for gemological testing

Francesco Poggialini, Beatrice Campanella, Stefano Legnaioli, Simona Raneri and Vincenzo Palleschi

Laser-induced breakdown spectroscopy (LIBS) is an analytical technique widely applied in many fields, including geological materials and Cultural Heritage. Its use in gemological analysis, although at the present rather limited, is very promising because of the intrinsic characteristics of the technique, which allow for a fast analysis on untreated samples at costs extremely competitive with respect to other analytical techniques commonly used in the field. Recent improvements of the LIBS technique strongly reduce the surface damages on the samples, thus making possible the analysis of precious stones and gems without leaving visible traces on the gemstones.

1.1 Introduction

The analysis of gemstones goes, nowadays, well beyond the classic visual inspection of the samples. The human experience is still essential, but in some cases, it must be complemented with technological tools which would help in assessing important issues such as authenticity, provenience, and genesis of the gemstones.

In the last 30 years, Raman spectroscopy [1] has successfully complemented refractive index and UV–visible spectroscopy analysis in gemology [2], being routinely used because of its non-destructive nature and the possibility to work on the sample without treatment of the surface, a fact that also allows for *in situ* applications of the technique using portable instrumentation [3]. X-ray elemental techniques such as x-ray fluorescence (XRF) and particle-induced x-ray emission (PIXE) have been successfully used for characterization of Cultural Heritage gemstones [4–6]; both of these techniques are not destructive, and several portable energy-dispersive x-ray instruments are commercially available, offering in some cases the possibility of obtaining elemental images of the gemstones [7].

Despite the availability of these fully non-destructive techniques, recently, minimally destructive techniques, based on the principle of laser ablation, have also received some attention. Among these, laser ablation inductively coupled plasma mass spectrometry (LA-ICP-MS) has been presented as a viable alternative in gemology to the non-invasive analytical methodologies as well as to other mass spectrometric techniques such as secondary ions mass spectrometry (SIMS) [8]. The rationale for using a minimally destructive technique for the analysis of precious stones, when fully non-destructive techniques are available, resides in the possibility of obtaining precise quantitative results on the concentration of key elements in the gems, potentially also in depth (up to several micrometers under the surface) by exploiting the mechanism of laser ablation. The high spatial resolution of the laser-based techniques guarantees in any case a minimal impact on the gem; moreover, the unavoidable removal of the material through the laser ablation mechanism can be exploited for marking a micro-logo on the gem surface to certify the testing of the gem [8].

Although LA-ICP-MS is, at present time, one of the most advanced analytical techniques for elemental analysis, its cost and the complexity of the analytical procedures make it unpractical to use as a routine technique for precious stones analysis. In fact, building multi-elemental calibration curves/surfaces is not easy in laser ablation–based techniques because of the lack of suitable standards which would cover all the possible variations that are encountered in gemology.

In the last few years, another laser ablation–based technique has raised the attention of the analytical chemistry community as a possible alternative to LA-ICP-MS. The name of the technique is laser-induced breakdown spectroscopy (LIBS), and its basic principles will be presented in the next section.

1.2 What is LIBS

LIBS is an analytical elemental technique which exploits the principle of laser ablation to remove a minimal amount of matter from the sample, as in LA-ICP-MS. However, while in LA-ICP-MS the ablated mass is further ionized through an ICP torch for undergoing mass spectrometry analysis, in LIBS, the ablated mass is already in a high-temperature, highly ionized status called plasma. The laser-induced plasma (LIP) is mainly formed by atoms, ions, and free electrons; a few microseconds after the ablation the temperature of the LIP can reach several thousands of degrees. In those conditions, atoms and ions are excited to higher energy levels and then decay back to the lower allowed energy level, emitting a characteristic radiation at a wavelength corresponding to the difference in energy between the upper and lower level of the transition [9].

If the radiative processes are slower than the characteristic time needed for the plasma to reach the thermal equilibrium, the plasma is said to be in local thermal equilibrium (LTE) [10, 11], and it can be described by two simple equations. The first is the Boltzmann equation:

$$n_k^{\mathrm{I/II}} = N^{\mathrm{I/II}} g_k \frac{e^{-\frac{E_k}{k_\mathrm{B}T}}}{U^{\mathrm{I/II}}(T)} \tag{1.1}$$

where n_k represents the number of atoms (n_k^{I}) or ions (n_k^{II}) of a given element whose electrons are at energy E_k, g_k is the degeneracy of the upper level, N is the total number of the corresponding species, $U(T)$ is the partition function of the species at the temperature T, and k_B is the Boltzmann constant. The Boltzmann equation (1.1) describes the equilibrium distribution of the bound electrons among the atomic and ionic energy levels at the temperature T.

The second equation describing plasma at LTE is the Saha–Eggart equation:

$$\frac{N^{\mathrm{II}}}{N^{\mathrm{I}}} = \frac{1}{n_\mathrm{e}} \frac{(2\,\pi\, m_\mathrm{e}\, k_\mathrm{B}T)^{3/2}}{h^3} \frac{2U^{\mathrm{II}}(T)}{U^{\mathrm{I}}(T)} e^{-\frac{E_{\mathrm{ion}}}{k_\mathrm{B}T}} \tag{1.2}$$

where h is the Planck constant, m_e is the electron mass, and n_e is the electron number density in the plasma.

The Saha–Eggart equation describes the equilibrium between two consecutive ionization stages of each element in the plasma. It is valid for any degree of ionization; however, in LIBS plasmas, only neutral and firstly ionized atoms are typically present, so E_{ion} in equation (1.2) represents the first ionization energy of the element.

For a transition involving two energy levels, with energies E_k and E_i, the total number of photons emitted per unit of time is proportional through a geometrical factor F to the concentration $C^{\mathrm{I/II}}$ of the atoms/ions in the upper level of the transition times the probability of spontaneous transition between the two levels:

$$I = F\, C^{\mathrm{I/II}} \frac{e^{-\frac{E_k}{k_\mathrm{B}T}}}{U^{\mathrm{I/II}}(T)} g_k A_{ki} \tag{1.3}$$

Equation (1.3) is the fundamental equation of LIBS analysis, since it links the total number of photons emitted per unit of time in a given transition to the number of the emitting atoms/ions. Since in LIBS the elements are present either in their neutral or singly ionized species, the higher ionization species being usually negligible, the total concentration of a given element can be written as follows:

$$C_{\mathrm{el}} \cong C^{\mathrm{I}} + C^{\mathrm{II}} = C_{\mathrm{I}}\left(1 + \frac{C^{\mathrm{II}}}{C^{\mathrm{I}}}\right) \tag{1.4}$$

Equation (1.4) allows for the determination of the element concentration in the LIBS plasma even when the emission lines of the ionized species are not visible/measurable in the spectrum, provided the plasma electron temperature T and number density n_e are known.

This consideration is at the basis of the calibration-free LIBS (CF-LIBS) method, invented by the Pisa group in 1999 [12]. CF-LIBS is a multi-elemental standard-less procedure that allows one to obtain precise quantitative results without the need for standard reference samples of known composition, using only the information contained in the LIBS spectrum and the knowledge of the fundamental spectroscopic parameters (transition probability and partition function) from the available atomic databases [13, 14].

For the application of the CF-LIBS procedure, the electron temperature and number density must be determined. Assuming the plasma is in LTE [10, 11], equation (1.3) can be rewritten, taking the logarithm of both the terms of the equation, as follows:

$$\log\left(\frac{I}{g_k A_{ki}}\right) = \log\left(\frac{F \quad C_s}{U(T)}\right) - \frac{E_k}{k_B T} \tag{1.5}$$

Equation (1.5) describes the relation between the measured intensity of an emission line and the concentration C_s of the corresponding emitting species. The relation can be put in a graphical form by plotting, for each emission line in the LIBS spectrum, the logarithm of $\left(\frac{I}{g_k A_{ki}}\right)$ as a function of the energy of the upper level of the transition (see figure 1.1).

The resulting plot is called Boltzmann plot, in which—in an ideal plasma in LTE—the emission lines of each species are aligned in parallel lines, with the same slope $\frac{1}{k_B T}$ and an intercept b_s that is proportional to the logarithm of the concentration of the species. Note that the plasma temperature, in the LTE approximation, is the same for all the species, i.e. it is a global parameter of the plasma. For its determination we need a minimum of two lines emitted by at least one species. All the other species will have the same temperature; therefore, it is

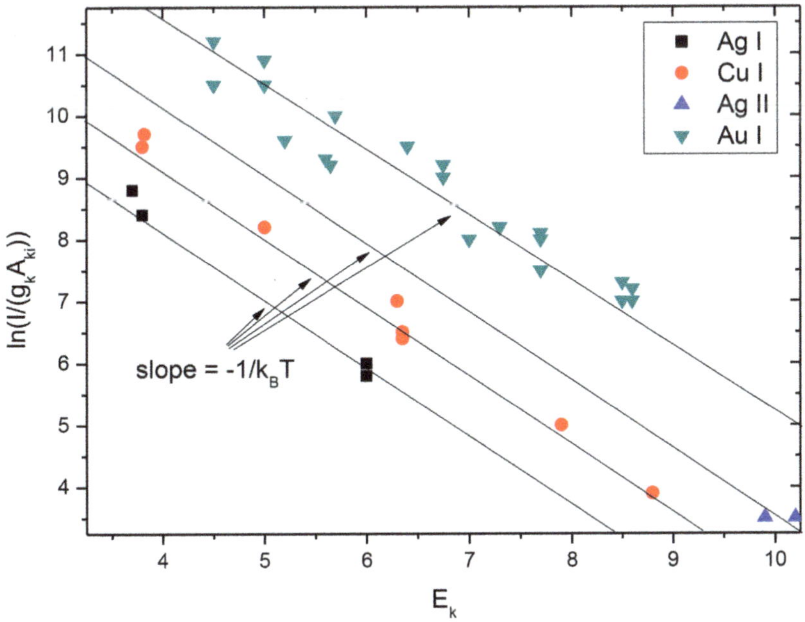

Figure 1.1. A typical Boltzmann plot (gold precious alloy).

possible to determine the intercept corresponding to species even when only a single emission line is visible in the LIBS spectrum.

Inverting the relation between the intercept and the species concentration, we obtain that

$$C_s = \frac{U(T) \quad e^{b_s}}{F} \tag{1.6}$$

If all the species of all the elements present in the plasma have at least one measurable line, the proportionality factor F can be determined by exploiting the fact that the sum of all the percent concentrations of the species in the plasma should correspond to 100%:

$$\sum C_s = \sum \frac{U(T) \quad e^{b_s}}{F} = 100\% \tag{1.7}$$

If only the emission of a neutral or ionized species of a given element is measurable in the LIBS spectrum, the concentration of the missing species can be determined using equations (1.2) and (1.4). For that, it would be necessary to determine the electron number density of the plasma, which is again a global parameter and can be determined by exploiting the link between the electron number density and the full-width at half-maximum (FWHM) of the emission lines. The fluctuating electric field generated by the free electrons around the atoms/ions produces a broadening of the emission lines through the Stark effect [15]. For non-hydrogenic lines, the Stark broadening is proportional to the electron number density through a coefficient (Stark coefficient) which is weakly dependent on the plasma electron temperature. The Stark broadening coefficients are tabulated for many emission lines; since the Stark broadening is typically dominant in LIBS plasmas, the determination of the electron number density is, in principle, relatively simple. However, the Stark broadening of non-hydrogenic lines is typically small, and it can be difficult to measure using low spectral resolution spectrometers. For that reason, it is sometimes preferable to determine the electron number density through the measurement of the Stark broadening of the hydrogen Balmer alpha line at 656.28 nm [16]; although the dependence of the hydrogen line FWHM is not linear with the electron number density, the problems of managing a non-linear dependence are compensated by the substantial independence of the measured FWHM on the spectrometer spectral resolution (see figure 1.2).

The hydrogen Balmer alpha line is always present in LIBS spectra obtained in air; although the source of hydrogen is the air humidity, it has been demonstrated that the hydrogen spectral signal has the same spatial distribution of the signal coming from the sample, while atmospheric nitrogen and oxygen show a different behaviour [17]. Therefore, the FWHM of the Balmer alpha depends on the electron number density of the bulk of the plasma, and it can be safely used for determining this parameter.

The CF-LIBS method has represented a revolution in LIBS quantitative analysis, offering a simple solution to the problem of the matrix effect in LIBS. The matrix

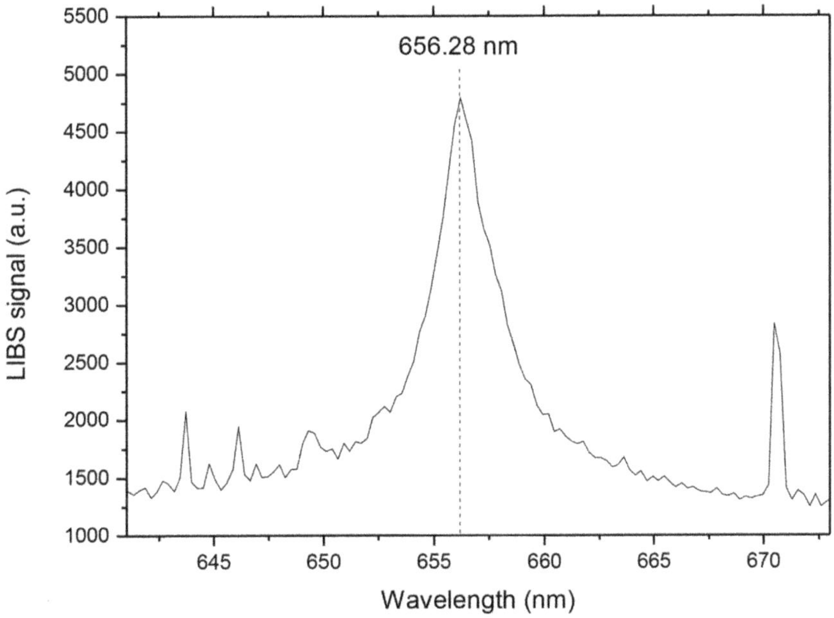

Figure 1.2. The hydrogen Balmer alpha line.

effect in LIBS makes it so that the spectral signal corresponding to the same concentration of a given element typically varies depending on the composition of the sample; this is due to the changes in the global parameters of the plasma (ablated mass, electron temperature, and number density) that may change with the change of the chemical–physical properties of the sample. In conventional laser-ablation quantitative measurements, based on the building of calibration curves, the matrix effect is overcome using matrix-matched standards (when available). On the other hand, CF-LIBS does not rely on the use of calibration standards because the global plasma parameters that might change, altering the sample matrix, are measured from the spectrum, and their effect on the signal is taken into account by the CF-LIBS algorithm. In this way, the LIBS technique gains a definite advantage with respect to other laser-ablation techniques, such as LA-ICP-MS, because it does not need calibration samples and does not suffer from matrix effects. In the last two decades, the CF-LIBS method has been substantially improved to take into account the effects associated with an optically thick plasma (part of the emitted radiation is re-absorbed by the plasma before reaching the spectrometer) [18, 19] or exploiting the benefits of artificial intelligence [20]. The experimental uncertainty on the spectroscopic parameters (transition probabilities and partition functions) can be almost completely reduced by using a single standard approach to CF-LIBS (one-point calibration—OPC) [21] or exploiting the changes in the line emission produced by changing the acquisition time after the laser pulse [22].

However, if the CF method gives LIBS an advantage with respect to LA-ICP-MS in analytical applications, it is true that the latter largely outperforms LIBS in terms of sensitivity and precision. To improve the analytical performance of LIBS, two important variations of the technique have been proposed: double-pulse LIBS (DP-LIBS) [23] and nanoparticle-enhanced LIBS (NELIBS) [24]. These two methods represent two completely different approaches to the issue of improving sensitivity and accuracy with respect to the rather poor analytical performances of the conventional approach to the LIBS technique. In DP-LIBS, the enhancement of the LIBS signal is obtained by focusing on the sample two laser pulses, usually collinear, with a suitable delay. The LIBS signal is collected after the second pulse. The physical principle of DP-LIBS was originally conjectured by Pershin in 1989 [25], but the fundamental role of the shock wave generated by the first laser pulse in reducing the plasma shielding of the second pulse was quantitatively demonstrated much later by Corsi *et al* [26, 27]. In DP-LIBS, no treatment of the sample is required. Therefore, DP-LIBS can be used at a distance in stand-off configuration [28]; the only constraint is the need for using two lasers, suitably synchronized, which in principle would increase the complexity and cost of a DP-LIBS instrument with respect to its conventional counterpart. However, double-pulse lasers are available on the market at costs comparable with single-pulse lasers.

In any case, if DP-LIBS can be considered a ‘physical’ approach to LIBS signal enhancing (no sample treatment, improving of the experimental setup to increase the signal), a few years ago, a ‘chemical’ alternative was proposed by De Giacomo *et al* [24], in which the experimental setup is not modified with respect to conventional LIBS, and the enhancement is obtained by treating the sample surface with metallic nanoparticles. This approach was called NELIBS; it exploits the local increase of the laser electric field produced by the nanoparticles (and the surface of the sample, when metals are analysed) to enhance the atomization of the ablated mass and, consequently, the LIBS signal. The physical mechanism underlying NELIBS has been recently discussed by De Giacomo *et al* [29]. The LIBS signal enhancements reported in recent NELIBS experiments [30–33] are rather impressive; however, it has been noted that the optimum signal is achieved using relatively high laser energies and at a given concentration of the nanoparticles on the sample surface. This latter parameter is difficult to control precisely, since the nanoparticles may aggregate on the sample surface, producing spatial inhomogeneities which may give problems, for example, in classification studies [34]. Further disadvantages of NELIBS are the need for treating the sample surface with chemical substances potentially dangerous for the environment and the constraint of close contact with the samples, which limits the use of NELIBS to laboratory analysis.

More recently, the Pisa group has proposed the combined use of DP-LIBS and NELIBS in a new LIBS approach called double-pulse NELIBS (DP-NELIBS) [35]. The authors demonstrated the possibility of obtaining a substantial enhancement of the DP-NELIBS signal with respect to DP-LIBS performed on untreated samples and conventional single-pulse NELIBS; to avoid the removal of the nanoparticles by the first laser pulse, the authors used an ingenious parallel, non-collinear configuration of the laser pulses [36] which allowed the generation of the shock wave by the first laser pulse without perturbing the interaction of the second pulse with the deposited nanoparticles on the sample surface (see figure 1.3).

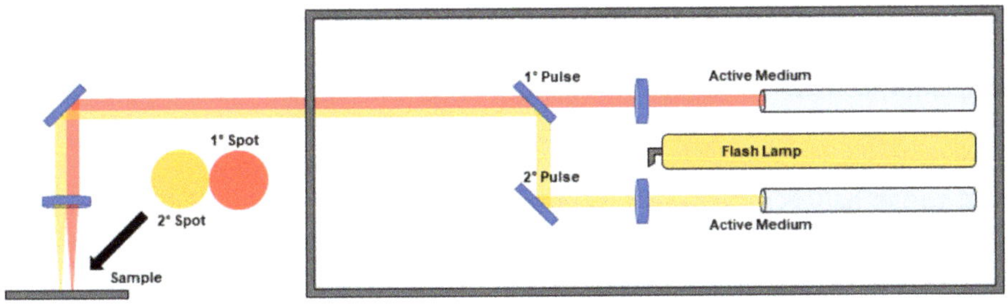

Figure 1.3. Experimental setup for DP-NELIBS.

Concluding this rapid overview of the LIBS technique, we can safely affirm that—at least for the specific application to gemology—LIBS offers the same capability of high spatial resolution, in-depth analysis, and mapping offered by LA-ICP-MS (the lasers used for the laser ablation in LA-ICP-MS are the same kind as the ones used in LIBS). From the point of view of quantitative analysis, the use of CF-LIBS allows for a standard-less analysis, overcoming the matrix effect which negatively affects the precision of LA-ICP-MS quantitative analysis. Finally, the use of DP-LIBS and/or NELIBS greatly improves the sensitivity of LIBS analysis, which is still poorer than LA-ICP-MS but nevertheless sufficient for providing useful quantitative information for assessing authenticity, provenience, and genesis of the gemstones, at a cost that can be substantially lower than an LA-ICP-MS apparatus [8].

1.3 Applications of LIBS in gemology

Despite the evident advantages of LIBS with respect to other laser-ablation techniques, its applications in gemology are still limited; below we will describe the most promising results reported in the literature.

In 2006, Abduriyim and Kitawaki made a comparison between the analytical capabilities of LA-ICP-MS, SIMS, and LIBS for detecting new treatments by the analysis of beryllium diffusion–treated corundum [8].

The authors used a conventional single-pulse LIBS system developed by SSEF Swiss Gemmological Laboratory [37], using a Nd:YAG pulsed laser (Big Sky Quantel) for ablation and generation of the plasma and an Ocean Optics LIBS2000 + wide-band spectrometer, covering the spectral range between 200 and 980 nm, for spectral analysis of the emitted radiation. Despite the relative simplicity of the LIBS setup, the authors reported limits of detection around 1–2 ppm for Be in corundum, which should be enough for identifying Be diffusion–treated corundum, which typically shows Be concentrations ranging from a few ppm to several hundred ppm. Considering the fact that the study reported in [8] was done with the LIBS technology of the early 2000s, it is reasonable to expect that using modern LIBS equipment would probably make feasible the study of Be at lower concentrations or, in any case, would improve the signal at the lower end of the concentration range of interest.

In 2008, McManus *et al* [38] analysed by LIBS 39 beryl samples from 11 mines in three USA states (New Hampshire, Connecticut, and Maine) to assess the capability of the technique in determining their provenience. The authors used a LIBS setup with a Q-switched Nd:YAG laser operating at the fundamental wavelength (1064 nm) coupled to an Ocan Optics broadband spectrometer with a resolution of 0.1 nm in the spectral range between 200 and 980 nm. They obtained some interesting results but admitted, also on the basis of their previous experience on LIBS analysis or 96 beryls from 16 countries [39], that a much larger dataset would be needed for a full successful classification. The authors also noted that the statistical method they used for classification (principal component analysis) [40] was not sensitive to the chromophores in the beryl, or, as they said, 'LIBS spectra appear to be color blind.' Therefore, their analysis was not able to distinguish between aquamarines (blue), goshenites (colorless), heliodors (yellow), or morganites (pink).

A few years after, in 2012, De Giacomo *et al* [41] described an application of CF-LIBS to the elemental analysis of an alexandrite gemstone (ideal formula $BeAl_2O_4$) from the Tokowaya River mines near Ekaterinburg, Russia. The authors use a single-pulse Nd:YAG laser in the second harmonics (532 nm) from Quanta System, Italy, for generating the plasma at the surface of the gemstone. The plasma emission light was then sent to a high spectral resolution monochromator (0.011 nm at 313 nm, the peak of the Be II emission) coupled to an intensified charge coupled device for the acquisition of the LIBS spectra. The authors made a systematic analysis of the plasma parameters, arriving to the conclusion that the plasma was in nearly LTE conditions in the temporal window between 1 and 6 μs after the laser pulse. In that interval they calculated the electron temperature of the plasma from the Boltzmann plot of 11 neutral iron emission lines; the electron number density was evaluated from the Stark broadening of one of the strong Al lines well visible in the LIBS spectrum.

Using the CF-LIBS algorithm, De Giacomo *et al* finally obtained the composition of the alexandrite gemstone (Be 7 ± 1 wt%, Al 41 ± 6 wt%, Fe 1.1 ± 0.1 wt%, Cr 0.22 ± 0.03 wt%). Since the measurements were performed in ambient air, the concentration of oxygen could be determined only by difference (50 ± 7 wt%). The results obtained were in good agreement with the concentrations obtained using a wavelength-dispersive x-ray spectroscopy (WDX) electron microprobe; the stoichiometric ratio between Al and Be was found to be 1.8 ± 0.3 using LIBS compared to 1.97 ± 0.03 from WDX. Similar good agreement was obtained for Cr/Al and Fe/Al stoichiometric ratios.

In 2014, Rossi *et al* [42] used LIBS to study the composition of gem-quality alexandrite, hiddenite, kunzite, elbaite, and topaz minerals. The authors concluded that the intrinsic multi-elemental nature of LIBS would be useful in gemological studies aimed at fast identification and characterization of the minerals through the determination of their composition, including chromophores and light lithophile elements (Li, Be, and B).

Another application of LIBS in gemology was presented in the same year by Agrosì *et al* [43]. The authors used a double-pulse micro-LIBS instrument in a study aimed to distinguish the manufacturing process of synthetic emeralds. For their study, Agrosì *et al* used a DP-LIBS instrument (Modì [44], Mobile Dual-Pulse Instrument by Marwan Technology, Pisa) composed by a double-pulse Q-switched Nd:YAG laser emitting at the fundamental wavelength (1064 nm) coupled with an Avantes Avaspec Dual-Channel Fiber Optic Spectrometer, covering the spectral range between 190 and 900 nm. The energy of the two laser pulses was set to 60 mJ with a delay of 2 μs between the first and second pulse. The laser beams were focused on the samples using a 10 cm focal length lens; the measurements were done in ambient atmosphere.

The LIBS analysis was performed on synthetic emeralds grown by flux method (Catham) or hydrothermal method (Biron) to evaluate the capability of LIBS to distinguish between emeralds grown by different methods and from different producers. The LIBS spectra were quantitatively analysed by applying the OPC [21] variation of the CF-LIBS method, using as a reference sample a Biron emerald previously characterized using LA-ICP-MS.

The main result reported in reference [43] was the capability of DP-LIBS + OPC analysis to perfectly distinguish the Catham from the Biron emeralds based on the concentration of the chromophore elements (Cr, Fe, V). As a further finding, the composition of the Catham samples was more homogeneous than the Biron one because the flux growth produces small homogeneous crystals compared to the hydrothermal method, in which the crystals are larger and more variable in composition.

Agrosì *et al* also noticed that, in a few cases, an imperfect focusing of the laser on the emerald surface would produce craters on the surface up to 200 μm in diameter. To overcome this problem, Tempesta and Agrosì [45] used the μ-Modì instrument [46], an evolution of Modì where the laser is coupled to a petrographic microscope for reducing the energy of the laser on the sample (and, consequently, the size of the laser crater) while maintaining a workable LIBS signal (see figure 1.4).

Using a microscope approach, the authors were able to reduce the size of the craters to under 10 μm; they studied six samples of red beryl, finding that two of them were pezzottaite. The main discriminating elements were found to be Li and Cs. The CF-LIBS analysis of the spectra allowed for the quantification of major and minor elements in the sample; the Mn concentration, in particular, was found to be in close correlation with the saturation of the red color in the samples.

A recent work from the same authors [47] demonstrated that DP-LIBS, coupled with a CF approach, would produce mineral formulas for red beryl compatible with the single-crystal structure refinement data from single-crystal x-ray diffraction analysis.

In 2015, Kochelek *et al* [48] demonstrated the possibility of classifying by LIBS a large quantity of rough sapphire and ruby specimens from 21 localities in 11 countries. In all the cases, the authors reported very high accuracy of the classification by country of origin (97.9% sapphire and 95.4% ruby) and by deposit of origin (98.9% sapphire and 96.0% ruby). For reaching this level of accuracy, they used a multivariate approach (partial least-squares regression) [49] to consider the complex compositional signature corresponding to the different deposits.

One of the most promising evolutions of LIBS analysis in gemology is represented by the work published in 2018 by the group of Alessandro De Giacomo [50].

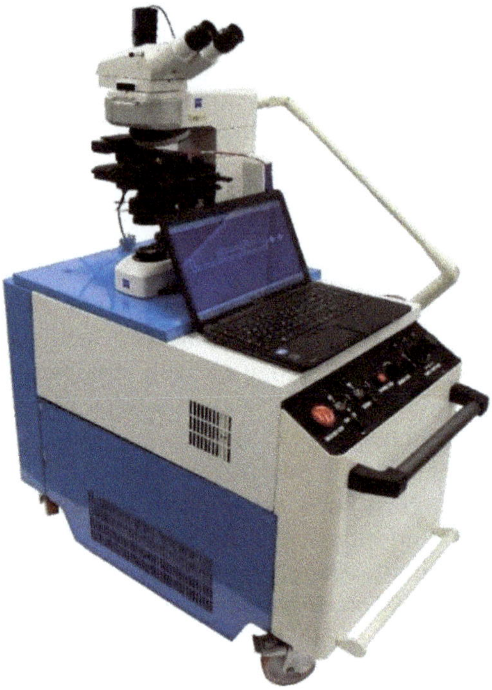

Figure 1.4. The μ-Modì DP-LIBS instrument.

The authors used a gold nanoparticle surface treatment for reducing the surface damage in LIBS analysis of diamonds, exploiting the principles of NELIBS on non-conducting media. When working on transparent samples, it might be important to 'shield' the surface to reduce the impact of the laser and protect the sample against excessive ablation; the nanoparticle surface layer therefore has the function to absorb efficiently the laser energy, preventing the direct ablation and producing a primary plasma that would bring to the plasma state only a few superficial layers of the sample, thus minimizing the visible effect of the ablation.

For their study, the authors used a Quantel Nd:YAG laser, emitting a pulse of energy up to 500 mJ in 6 ns at 532 nm. The laser spot was defocused on the gem surface to about 1.5 mm in diameter. The LIBS signal was collected by a lens on a Czerny–Turner monochromator coupled to an intensified charge-coupled device (CCD) camera. The Au nanoparticles were generated by pulsed laser ablation in liquid and had a diameter of 15 ± 2 nm. In previous works [51], De Giacomo *et al* have shown that in NE-LIBS application to non-conducting samples, the ablation occurs at the spot of the nanoparticles and the size of the nano-craters is the same as the nanoparticles, i.e. completely invisible to the naked eye.

The authors of [50] remarked that, in the application to gemology, one should not expect a substantial decrease of the limits of detection in the ppm range for most of the elements of interest, but these limits are achieved using a laser energy much lower with respect to standard LIBS without producing any visible damage on the gem surface. The group of De Giacomo studied tourmaline, aquamarine, and ruby samples; in all cases they were able to measure major and minor elements in the samples without producing any visible damage on the gemstone surface. Moreover, using CF-LIBS they also determined the atoms per formula unit of the main tourmaline elements (Al, Si, and B), finding a good agreement with the results of LA-ICP-MS [42].

Despite the excellent capability of LIBS for classifying gemstones according to their provenience and genesis, in some cases satisfactory results cannot be obtained using this technique alone. In a recent paper, Eum *et al* [52] demonstrated that raw sapphires coming from different countries (Laos, Rwanda, Congo, Mozambique, and Australia) can be classified according to their geographical origin with >90% precision using a combined Raman/LIBS approach. The LIBS spectra were collected using a J200 Tandem instrument from Applied Spectra, equipped with a Nd:YAG laser at 266 nm and a wide-band spectrometer covering the spectral range between 185 and 1048 nm, with a spectral resolution of 0.1 nm. The study of the authors reported an accuracy of Raman classification alone of 84.4% and LIBS classification alone of 81.3%; however, when merging the two, the overall accuracy reached 87.5%. The use of a statistical technique called hierarchical support vector machine allowed for a selective approach, adopting only Raman or LIBS at different steps of the classification. In this way, the accuracy of the classification increased to 90.6%.

A similar classification study was recently published by Dubey *et al* [53], which used CF-LIBS for studying the composition of amethyst, aquamarine beryl, bloodstone citrine, diopside, and enstatite. The peculiarity of their work was the use of LIBS for (indirectly) measuring the hardness of the gemstones; several works in the literature demonstrate that a correlation exists between the sample hardness and the LIBS plasma parameters (electron number density and electron temperature) [54–57]. The authors found a good linear correlation between electron temperature and Mohs hardness of the gemstones, with the higher plasma temperature corresponding to the higher hardness. The authors observed a variation in the plasma temperature of about 20% when passing from a Mohs hardness of 5.5 (enstatite) to a hardness of 7.5 (aquamarine beryl).

1.4 Conclusion

There are many analytical techniques that can be applied for assessing authenticity, provenience, and genesis of gemstones. None of them is free from drawbacks, so in some cases, as we have seen, several independent techniques must be used together for improving the accuracy of the results obtained. In this complex scenario, the LIBS technique represents a now consolidated method, competitive with LA-ICP-MS in terms of performance and cost. The clear advantages of LIBS with respect to other laser-ablation techniques are the possibility of performing an accurate quantitative analysis of the gemstones without using calibration curves, using the CF-LIBS technique, and the possibility of eliminating any visible damage on the gemstone using a nanoparticles coating on the surface irradiated by the laser. These two peculiar characteristics open the way to a wider application of the LIBS technique in gemology. The possible combination of double-pulse and NELIBS should also be explored, since it might offer the possibility of performing accurate quantitative LIBS analysis of gemstones without producing surface damage, as occurs in LA-ICP-MS.

References

[1] Kiefert L and Karampelas S 2011 Use of the Raman spectrometer in gemmological laboratories: review *Spectrochim. Acta* A **80** 119–24

[2] Hainschwang T 2017 Gemstone analysis by spectroscopy *Encyclopedia of Spectroscopy and Spectrometry* (Academic Press) pp 18–24

[3] Bersani D and Lottici P P 2010 Applications of Raman spectroscopy to gemology *Anal. Bioanal. Chem.* **397** 2631–46

[4] Barone G, Mazzoleni P, Bersani D and Raneri S 2017 Portable XRF: a tool for the study of corundum gems *Open Archaeol.* **3** 194–201

[5] Pappalardo L, Karydas A G, Kotzamani N, Pappalardo G, Romano F P and Zarkadas C 2005 Complementary use of PIXE-alpha and XRF portable systems for the non-destructive and *in situ* characterization of Gemstones in Museums *Nucl. Instrum. Methods Phys. Res.* B **239** 114–21

[6] Karydas A G, Kotzamani D, Bernard R, Barrandon J N and Zarkadas C 2004 A compositional study of a museum jewellery collection (7th–1st BC) by means of a portable XRF spectrometer *Nucl. Instrum. Methods Phys. Res.* B **226** 15–28

[7] Lu Z, He X, Lin C, Liang L, Jin X and Guo Q 2020 Color and genesis of californite from Pakistan: insights from μ-XRF mapping, optical spectra and x-ray photoelectron spectroscopy *Sci. Rep.* **10** 285

[8] Abduriyim A and Kitawaki H 2006 Applications of laser ablation–inductively coupled plasma-mass spectrometry (LA-ICP-MS) to gemology *Gems Gemol.* **42** 98–118

[9] Palleschi V 2020 Laser-induced breakdown spectroscopy: principles of the technique and future trends *ChemTexts* **6** 18

[10] Fujimoto T and McWhirter R W P 1990 Validity criteria for local thermodynamic equilibrium in plasma spectroscopy *Phys. Rev.* A **42** 6588–601

[11] Cristoforetti G, De Giacomo A, Dell'Aglio M, Legnaioli S, Tognoni E, Palleschi V and Omenetto N 2010 Local thermodynamic equilibrium in laser-induced breakdown spectroscopy: beyond the McWhirter criterion *Spectrochim. Acta* B **65** 86–95

[12] Ciucci A, Corsi M, Palleschi V, Rastelli S, Salvetti A and Tognoni E 1999 New procedure for quantitative elemental analysis by laser-induced plasma spectroscopy *Appl. Spectrosc.* **53** 960–4

[13] NIST https://physics.nist.gov/PhysRefData/ASD/lines_form.html

[14] https://cfa.harvard.edu/amp/ampdata/kurucz23/sekur.html

[15] Griem H R 1976 Stark broadening *Adv. At. Mol. Phys.* **11** 331–59

[16] Pardini L, Legnaioli S, Lorenzetti G, Palleschi V, Gaudiuso R, De Giacomo A, Diaz Pace D M, Anabitarte Garcia F, De Holanda Cavalcanti G and Parigger C 2013 On the determination of plasma electron number density from stark broadened hydrogen balmer series lines in laser-induced breakdown spectroscopy experiments *Spectrochim. Acta* B **88** 98–103

[17] De Giacomo A, Dell'Aglio M, Gaudiuso R, Cristoforetti G, Legnaioli S, Palleschi V and Tognoni E 2008 Spatial distribution of hydrogen and other emitters in aluminum

laser-induced plasma in air and consequences on spatially integrated laser-induced breakdown Spectroscopy measurements *Spectrochim. Acta* B **63** 980–87

[18] Bulajic D, Corsi M, Cristoforetti G, Legnaioli S, Palleschi V, Salvetti A and Tognoni E 2002 A procedure for correcting self-absorption in calibration free-laser induced breakdown spectroscopy *Spectrochim. Acta* B **57** 339–53

[19] Safi A, Tavassoli S H H, Cristoforetti G, Tognoni E, Campanella B, Legnaioli S, Pagnotta S, Poggialini F and Palleschi V 2019 Exploiting self-absorption for plasma characterization in laser-induced breakdown spectroscopy experiments: a comparison of two recent approaches *Anal. Chem.* **91** 8595–601

[20] D'Andrea E *et al* 2015 A Hybrid calibration-free/artificial neural networks approach to the quantitative analysis of LIBS spectra *Appl. Phys.* B **118** 353–60

[21] Cavalcanti G H, Teixeira D V, Legnaioli S, Lorenzetti G, Pardini L and Palleschi V 2013 One-point calibration for calibration-free laser-induced breakdown spectroscopy quantitative analysis *Spectrochim. Acta* B **87** 51–6

[22] Urbina I, Carneiro D, Rocha S, Farias E E, Bredice F and Palleschi V 2020 Study of binary lead–tin alloys using a new procedure based on calibration-free laser induced breakdown spectroscopy *Spectrochim. Acta* B **170** 105902

[23] Tognoni E and Cristoforetti G 2014 Basic mechanisms of signal enhancement in Ns double-pulse laser-induced breakdown spectroscopy in a gas environment *J. Anal. At. Spectrom.* **29** 1318–38

[24] De Giacomo A, Gaudiuso R, Koral C, Dell'Aglio M and De Pascale O 2013 Nanoparticle-enhanced laser-induced breakdown spectroscopy of metallic samples *Anal. Chem.* **85** 10180–7

[25] Pershin S 1989 Transformation of the optical spectrum of a laser plasma when a surface is irradiated with a double pulse *Sov. J. Quantum Electron.* **19** 1618–9

[26] Corsi M, Cristoforetti G, Giuffrida M, Hidalgo M, Legnaioli S, Palleschi V, Salvetti A, Tognoni E and Vallebona C 2004 Three-dimensional analysis of laser induced plasmas in single and double pulse configuration *Spectrochim. Acta* B **59** 723–35

[27] Wen S-B *et al* 2005 Comment on 'three-dimensional analysis of laser induced plasmas in single and double pulse configuration *Spectrochim. Acta* B **60** 870–2

[28] Lorenzetti G, Legnaioli S, Grifoni E, Pagnotta S and Palleschi V 2015 Laser-based continuous monitoring and resolution of steel grades in sequence casting machines *Spectrochimica Acta* B **112** 1–5

[29] De Giacomo A, Salajkova Z and Dell'aglio M 2019 A quantum chemistry approach based on the analogy with π-system in polymers for a rapid estimation of the resonance wavelength of nanoparticle systems *Nanomaterials* **9** 1–16

[30] Koral C, De Giacomo A, Mao X, Zorba V and Russo R E 2016 Nanoparticle enhanced laser induced breakdown spectroscopy for improving the detection of molecular bands *Spectrochim. Acta* B **125** 11–7

[31] De Giacomo A, Koral C, Valenza G, Gaudiuso R and Dellaglio M 2016 Nanoparticle enhanced laser-induced breakdown spectroscopy for microdrop analysis at subppm level *Anal. Chem.* **88** 5251–7

[32] De Giacomo A, Dell'Aglio M, Gaudiuso R, Koral C and Valenza G 2016 Perspective on the use of nanoparticles to improve LIBS analytical performance: nanoparticle enhanced laser induced breakdown spectroscopy (NELIBS) *J. Anal. At. Spectrom.* **31** 1566–73

[33] Dell'Aglio M, Alrifai R and De Giacomo A 2018 Nanoparticle enhanced laser induced breakdown spectroscopy (NELIBS), a first review *Spectrochim. Acta* B **148** 105–12

[34] El-Saeid R H H, Abdel-Salam Z, Pagnotta S, Palleschi V and Harith M A A 2019 Classification of sedimentary and igneous rocks by laser induced breakdown spectroscopy and nanoparticle-enhanced laser induced breakdown spectroscopy combined with principal component analysis and graph theory *Spectrochim. Acta* B **158** 105622

[35] Poggialini F, Campanella B, Legnaioli S, Pagnotta S and Palleschi V 2020 Investigating double pulse nanoparticle enhanced laser induced breakdown spectroscopy *Spectrochim. Acta* B **167** 105845

[36] Cristoforetti G, Legnaioli S, Palleschi V, Tognoni E and Benedetti P A A 2010 Crater drilling enhancement obtained in parallel non-collinear double-pulse laser ablation *Appl. Phys.* A **98** 219–25

[37] Krzemnicki M S, Hänni H A and Walters R A 2004 A new method for detecting Be diffusion-treated sapphires: laser-induced breakdown spectroscopy (LIBS) *Gems Gemol.* **40** 314–22

[38] McManus C E, McMillan N J, Harmon R S, Whitmore R C, De Lucia F C Jr and Miziolek A W 2008 Use of laser induced breakdown spectroscopy in the determination of gem provenance: beryls *Appl. Opt.* **47** G72

[39] McMillan N J, McManus C E, Harmon R S, De Lucia F C and Miziolek A W 2006 Laser-induced breakdown spectroscopy analysis of complex silicate minerals—beryl *Anal. Bioanal. Chem.* **385** 263–71

[40] Pořízka P, Klus J, Képeš E, Prochazka D, Hahn D W and Kaiser J 2018 On the utilization of principal component analysis in laser-induced breakdown spectroscopy data analysis, a review *Spectrochim. Acta* B **148** 65–82

[41] De Giacomo A, Dell'Aglio M, Gaudiuso R, Santagata A, Senesi G S, Rossi M, Ghiara M R, Capitelli F and De Pascale O 2012 A laser induced breakdown spectroscopy application based on local thermodynamic equilibrium assumption for the elemental analysis of alexandrite gemstone and copper-based alloys *Chem. Phys.* **398** 233–38

[42] Rossi M, Dell'Aglio M, De Giacomo A, Gaudiuso R, Senesi G S, De Pascale O, Capitelli F, Nestola F and Ghiara M R 2014 Multi-methodological investigation of kunzite, hiddenite, alexandrite, elbaite and topaz, based on laser-induced breakdown spectroscopy and conventional analytical techniques for supporting mineralogical characterization *Phys. Chem. Miner.* **41** 127–40

[43] Agrosì G *et al* 2014 Application of laser induced breakdown spectroscopy to the identification of emeralds from different synthetic processes *Spectrochim. Acta* B **102** 48–51

[44] Bertolini A *et al* 2006 Modì: a new mobile instrument for *in situ* double-pulse LIBS analysis *Anal. Bioanal. Chem.* **385** 240–47

[45] Tempesta G and Agrosì G 2016 Standardless, minimally destructive chemical analysis of red beryls by means of laser induced breakdown spectroscopy *Eur. J. Mineral.* **28** 571–80

[46] Schiavo C, Menichetti L, Grifoni E, Legnaioli S, Lorenzetti G, Poggialini F, Pagnotta S and Palleschi V 2016 High-resolution three-dimensional compositional imaging by double-pulse laser-induced breakdown spectroscopy *J. Instrum.* **11** C08002

[47] Tempesta G, Bosi F and Agrosì G 2020 Crystal chemical characterisation of red beryl by 'standardless' laser-induced breakdown spectroscopy and single-crystal refinement by x-ray diffraction: an example of validation of an innovative method for the chemical analysis of minerals *Geostand. Geoanal. Res.* **44** 685–93

[48] Kochelek K A, McMillan N J, McManus C E and Daniel D L 2015 Provenance determination of sapphires and rubies using laser-induced breakdown spectroscopy and multivariate analysis *Am. Mineral.* **100** 1921–31

[49] Gottfried J L, Harmon R S, De Lucia F C and Miziolek A W 2009 Multivariate analysis of laser-induced breakdown spectroscopy chemical signatures for geomaterial classification *Spectrochim. Acta* B **64** 1009–19

[50] Koral C, Dell'Aglio M, Gaudiuso R, Alrifai R, Torelli M and De Giacomo A 2018 Nanoparticle-enhanced laser induced breakdown spectroscopy for the noninvasive analysis of transparent samples and gemstones *Talanta* **182** 253–8

[51] De Giacomo A, Gaudiuso R, Koral C, Dell'Aglio M and De Pascale O 2014 Nanoparticle enhanced laser induced breakdown spectroscopy: effect of nanoparticles deposited on sample surface on laser ablation and plasma emission *Spectrochim. Acta* B **98** 19–27

[52] Eum C, Jang D, Kim J, Choi S, Cha K and Chung H 2018 Improving the accuracy of spectroscopic identification of geographical origins of agricultural samples through cooperative combination of near-infrared and laser-induced breakdown spectroscopy *Spectrochim. Acta* B **149** 281–7

[53] Dubey S, Kumar R, Rai A K, Pati J K, Kiefer J and Rai A K 2021 Rapid analysis of chemical composition and physical properties of gemstones using LIBS and chemometric technique *Appl. Sci.* **11** 6156

[54] Abdel-Salam Z A, Nanjing Z, Anglos D and Harith M A 2009 Effect of experimental conditions on surface hardness measurements of calcified tissues via LIBS *Appl. Phys.* B **94** 141–7

[55] Abdel-Salam Z A, Galmed A H, Tognoni E and Harith M A 2007 Estimation of calcified tissues hardness via calcium and magnesium ionic to atomic line intensity ratio in laser induced breakdown spectra *Spectrochim. Acta* B **62** 1343–7

[56] ElFaham M M, Alnozahy A M and Ashmawy A 2018 Comparative study of LIBS and mechanically evaluated hardness of graphite/ rubber composites *Mater. Chem. Phys.* **207** 30–5

[57] ElFaham M M, Okil M and Mostafa A M 2018 Limit of detection and hardness evaluation of some steel alloys utilizing optical emission spectroscopic techniques *Opt. Laser Technol.* **108** 634–41

Chapter 2

Raman spectroscopy for the non-destructive analysis of gemstones

Danilo Bersani, Laura Fornasini, Peter Vandenabeele and Anastasia Rousaki

Raman spectroscopy is an ideal tool for the analysis of inorganic and organic materials in solid, liquid, or even gas form. The technique shares some favourable characteristics for the investigation of samples or specimens such as excellent spatial and spectral resolutions, fast identification, and non-destructive analysis. Moreover, one of the most noteworthy properties of the technique is that it can be applied directly and on site when using dedicated mobile Raman spectrometers. All the aforementioned advantages, together with the high structural information retrieved from Raman spectroscopy analysis, make the technique well suited for research on gemological materials. Thus, this chapter serves as a compact overview of the theory and applications of Raman spectroscopy in and out the laboratory in the case of gemstones, including pearls and corals. Moreover, selected examples of Raman spectroscopy applications are presented, underlining the effectiveness of the technique.

2.1 Raman spectroscopy on gemstones

Gem analysis is often used to produce clear answers to specific questions. The fundamental question is the value of the gem, and the first necessary information is then the mineralogical species of the gem: not only its chemical composition but also its structure. The second question is then the gem's origin: where does it come from (provenance), and is it natural or man-made? The identification of its geographical provenance helps to better define its value. And a clue to its geological history is useful for the mineralogists studying the gem's origin. Again, for natural gemstones it is important to identify the presence of undeclared enhancing treatments, such as oiling, fissure filling, dyeing, heat, or radiation treatments. All these questions should be answered without any damage to the gems, using a technique able to study even gems mounted in precious jewels or invaluable artefacts. As we will see in the

doi:10.1088/978-0-7503-3927-8ch2

following, Raman spectroscopy can contribute to many of these research questions in a fast and completely non-destructive way.

2.1.1 A short introduction to Raman spectroscopy

Different analytical approaches are used to examine the composition of minerals and rocks. When analysing gemstones, the list of available techniques is more limited, as the artefact has to be analysed in a non-destructive way, while the precious stone is often fixed in a jewel. Research questions usually involve the identification of the stone, the study of its treatments, or an investigation of the provenance of the gem. One approach that is perfectly fit for this purpose, is Raman spectroscopy, a molecular spectroscopic technique that is used to study vibrational, rotational, and other low-frequency modes in a system. The Raman effect was discovered in 1928 by Sir C V Raman, and the technique is now widely developed and exploited in various fields and applications, for example in pharmaceutical analysis, forensics, archaeometry, solid-state physics, and geology.

The Raman effect can be explained as a light scattering effect. It is produced by the interaction of light with atoms bound together. In the following we will refer to the group of atoms as molecules, but they can be also represented as crystalline or amorphous solids. A monochromatic (laser) source is focused on a molecule, and the scattered light is studied. It is considered that, at absorption, the molecule is excited to an unstable—virtual—state, and immediately it relaxes by emitting the energy in the form of a photon. If the energy of the absorbed photon equals the energy of the emitted photon, elastic scattering occurs. Almost all photons undergo this Rayleigh scattering. However, if the scattered photon has a different energy in comparison to the incident beam, the collision is said to be inelastic: there is an energy transfer between the molecule and the photon. This is also called Raman scattering, and two possible phenomena can be observed. If the molecule is promoted from the vibrational ground state to a virtual state, but relaxes to the first excited vibrational state, Stokes Raman scattering occurs. In this case the emitted photon has a lower energy than the incident photon. However, it may happen that the molecule is already excited before the Raman effect happens: if the molecule is transferred from the first vibrational excited state to the virtual state and relaxes to the ground state, then the scattered photon has more energy than the initial monochromatic light. This process is known as anti-Stokes Raman scattering. In a Raman spectrum, typically the energy difference between the scattered radiation and the incident laser beam (expressed as Raman wavenumbers, units: cm^{-1}) is plotted on the horizontal axis, while the vertical axis is an intensity scale. At 0 cm^{-1} energy difference a very intense band is observed, corresponding with the Rayleigh radiation. Usually, this band is suppressed by appropriate filtering, as it would overwhelm the much weaker Raman scattering. Stokes and anti-Stokes Raman bands are observed symmetrically around the Rayleigh line. Their position relates to the energy difference between the vibrational ground state and the first excited vibrational state, but they have a different sign, as in the Stokes case there is a net energy transfer from the laser beam towards the molecule (or the lattice of the crystal), and in the anti-Stokes case the net energy transfer is equal in size but goes from the molecule towards the scattered photon. Therefore, basically Stokes and anti-Stokes Raman bands contain the same information. As at thermal equilibrium, according to the Boltzmann distribution, the number of molecules in a lower state is always higher than the number of molecules in a higher vibrational state, Stokes scattering is typically more intense than anti-Stokes radiation. Therefore, one only reports the Stokes part of the spectrum. In order for the Raman effect to occur, the main selection rule has to be obeyed: during the vibration, a change in polarisability should occur. The polarisability is a measure that reflects how easily the electron cloud of a molecule can be distorted by an electromagnetic field.

Raman spectroscopy has many advantages in materials characterisation: there is almost no sample preparation needed, it is a relatively fast method, and the technique is able to deal with a broad range of materials. The laser power may be lowered by density filters to avoid any undesirable alteration of the sample due to heating effects. The method is contactless and non-destructive. Moreover, the technique can be coupled with different optics, like microscopes and fibre-optics probe heads. The latter, combined with a trend towards miniaturisation of the instrumentation, makes it possible to perform *in situ* measurements. Despite these advantageous properties, Raman spectroscopy suffers some drawbacks. From physics, we know that the Raman effect is relatively weak: it is prone to interferences that may overwhelm the spectrum. In particular, fluorescence (due to electronic transitions) causes a broad signal that is much stronger than the Raman effect. One possible approach to avoid this is shifting to different excitation wavelengths. Moreover, as Raman spectroscopy is a molecular spectroscopic technique, it is able to get information from molecules, crystals, and amorphous solids but not from isolated atoms. Even ionic solids (like salts) and metals produce little or no Raman signal. In that case, often Raman spectroscopy is combined with elemental spectroscopic approaches, such as x-ray fluorescence analysis.

2.2 Benchtop and mobile Raman spectroscopy

Before describing the types of Raman instrumentation, it is useful to understand how the technique can be applied on a material or an object. In general, Raman spectroscopy is a non-destructive technique. Therefore, the laser power should be kept sufficiently low in order not to introduce thermal degradation or structural alteration of our sample or object. Otherwise, the technique can be invasive, as samples can be extracted for measurements. Typically, micro-Raman laboratory systems involve invasive analysis on collected samples or on small objects able to fit in the system's chamber. The collected samples can be measured untreated or after manipulation, e.g. embedded in appropriate resin for stratigraphic analysis. Mobile Raman systems can also perform analysis on smaller or larger samples. Non-invasive analysis is conducted when the object under analysis is measured without sampling. This can be performed by both benchtop and mobile Raman systems. The terms direct, *in situ*, on site, on field, etc refer to the ability to study a material or an object in its natural or post-depositional environment.

Benchtop Raman spectrometers are the type of instrumentation typically used for laboratory experiments. These heavy but very stable systems were the first to emerge in the market years before mobile Raman spectrometers became available for the non-invasive and direct analysis of objects. More than 40 years have passed since the 'MOLE' Raman laser microprobe became commercially available from Jobin Yvon [1], a breakthrough in the Raman molecular analysis performed in the laboratory. This breakthrough quickly affected the type of analysis of a broad range of materials including gemstones and materials for geological purposes. This is evident by the examples described in Dhamelincourt in 1987 [1] and Coupry and Brissaud in 1996 [2]. Today's benchtop Raman systems are upgraded state-of-the-art instruments, using exceptional hardware and software design, simplifying the analysis in terms of operation and simultaneously producing reliable outcomes.

Nowadays, most of the Raman laboratory systems can be coupled with a wide selection of lasers, ranging from the deep ultraviolet (UV) to the near infrared (IR) region. They are accompanied with the appropriate detector according to the laser, e.g. charge-coupled devices, solid state (semiconductor), etc. Choosing the appropriate laser for analysis is a crucial step and completely depends on the material measured and the rules of spectroscopy that govern the laser-material system and its interactions. Shorter wavelengths are usually used for provoking resonance Raman effects. If the laser's energy is sufficiently high to excite the molecule in an electronic state, then the Raman signal can be amplified several orders of magnitude. Indeed, Bergamonti *et al* in 2011 [3] studied the application of a 473.1 nm laser on corals and pearls. A challenge that can be addressed when selecting short laser wavelengths is to avoid the fluorescence that might overwhelm the Raman spectrum. In this case, lasers in the near-IR region of the electromagnetic spectrum should be considered for suppressing fluorescence. But by shifting to near-IR lasers, the Raman cross section evidently decreases, and as a consequence higher energy powers and overall measuring time should be adopted to compensate for the low Raman signal collection.

Typically, in the past, for near-IR lasers, for example the 1064 nm laser, Fourier transform (FT) Raman spectroscopy was used. As a consequence, this type of spectrometer was coupled to a liquid nitrogen cooled detector. As the technology evolved through the years, the FT-Raman systems were replaced with 1064 nm dispersive Raman systems, using semiconductor thermoelectrically cooled detectors.

The high wavelength stability and the outstanding spatial resolution (affected by both the numerical aperture of the objective and the laser wavelength) and spectral resolution (affected by the diffraction grating, the focal length, the excitation wavelength, the detector, and the slit or pinhole of the spectrometer) of the benchtop Raman systems are some of the preferable characteristics when analysing several materials including gemstones, pearls, corals, etc.

Raman instruments for laboratory use are often coupled to variable magnification objectives with high magnifications accounting for short working distances. Several measuring spot sizes can be achieved by using different objectives. Indeed, some of the benchtop Raman spectrometers can reduce the spot size of analysis to just a few micrometres. Other features that can be controlled completely by the operator are the laser power (for not altering or thermally degrading the material) and the total measuring time. In confocal Raman systems the pinhole or slit can be also changed, thus decreasing or increasing the depth and lateral resolution. When increasing the size of the pinhole the depth and lateral resolution decreases and vice versa (when decreasing its size), with direct influence on the Raman signal collected.

One of the most undeniable advantages of the Raman technique is its application on site and directly on the object under analysis. For the purpose of direct analysis several milestones were reached, for example, in the field of cultural heritage studies. Micro-Raman systems were used for conducting research directly on the artefact [4–6]. Then, another option for direct analysis explored the possibility of attaching fibre optics onto a laboratory spectrometer (FT-Raman system) [7]. Finally, in 2004 the first compact, completely mobile Raman system was designed for the investigation of cultural heritage objects [8].

Several review studies based on mobile Raman spectroscopy (with or without reviewing laboratory and micro-Raman applications) emerged in the literature covering a broad range of fields including gemology [9], art, archaeology, archaeometry, geosciences, forensics [10], minerals found on works of art [11], cultural heritage studies [12, 13], geosciences and art analysis [14].

Raman spectrometers used for on-site measurements are not exclusively compact or mobile. According to Vandenabeele *et al* in 2014 [10], 'traditional' heavy-weighted Raman spectrometers can be carried out of the laboratory and thus are *transportable*. Of course, the whole packaging of such instrumentation is not specifically flexible, and some precautions should be taken regarding instrumental components, such as the optics. For the Raman systems that are designed for mobility (stable, scaled down in size, etc), one can use the umbrella term of *mobile* [10]. Then the mobile spectrometers can be further differentiated to portable, handheld, and palm sized [10]. *Portable* Raman systems use fibre optics for the laser and the signal collection. They are usually compact, work on battery, and can be carried and operated by a single person [10]. Moreover, their laser power and total measuring time can be adjusted by the operator according to the needs of analysis. *Handheld* Raman spectrometers use fixed optical heads attached to their light-weight main unit, while some of their measuring conditions can be partially or totally changed by the user [10]. *Palm* instrumentation is further downscaled in size [10].

Although mobile Raman spectrometers are the ideal tool for fast, direct, and on-site analysis, one should be aware of several challenges regarding their application. System calibration is one of the most important aspects when measuring on the field. Mobile Raman systems might suffer wavelength instabilities due to transportation, environmental conditions, etc. Therefore, frequent calibration should be carried out for correctly identifying the unknown. When meticulous calibration is not possible the system should be frequently checked for wavenumber stability by measuring pure products with a known Raman band position.

Mobile Raman spectrometers generally use larger spot sizes of analysis and have poorer spectral resolutions. Today, this is not entirely the case, as mobile Raman systems can be attached to magnification objectives, thus achieving smaller measuring volumes while benefiting from higher spectral resolutions. For most of the mobile systems, features like laser power and total measuring time are completely adjustable by the user.

Positioning mobile Raman equipment that is attached to fibre-optics cables can be done either by using positioning accessories, e.g. tripods, or by holding the probe head by hand. Positioning accessories will not only securely place the probe head perpendicular to the measuring surface but also help its free movement on the x-, y-, and z-axes. Moreover, the stable placement of the probe head relative to the object will help focus the laser beam on the artefact. If the positioning is done by hand, in most cases special cups are slid over the probe head's lenses to simultaneously block the environmental signal and secure an optimal focal length (depending on the lens used) for acquiring measurements. With this, the probe head or the protective cup should come in complete contact with the surface under study. In general, light interferences should be avoided.

For Raman instruments that use a fixed optical head the measuring procedure is less complex. The contact lens of the optical head can come in contact with the desired object, and by securing a dark environment, the measurement can be conducted.

Usually, mobile Raman spectrometers are coupled to single lasers for analysis, although there are commercially available multiple laser systems. Moreover, a dual laser handheld Raman instrument was designed using a special patented technology for eliminating spectral artefacts, such as fluorescence [15]. Jehlička *et al* in 2017 [16] evaluated the use of seven commercially available mobile Raman spectrometers, with different characteristics and hardware and software technologies, when investigating beryls. Culka and Jehlička in 2019 [17, 18] used the dual laser handheld Raman instrument, with a sequentially shifted excitation algorithm for fluorescence suppression, to study several gemstones and minerals.

The possibilities of both benchtop micro-Raman and mobile Raman systems when applied on objects of direct gemological interest are remarkable. Bersani and Lottici in 2010 [9] gave an overview of Raman spectroscopy applied in gemology studies with laboratory and mobile Raman systems, while Coccato *et al* in 2021 [19] used laboratory micro- and portable Raman equipment (among other techniques) to study zoisite and tanzanite. The significance of the technique for gemological purposes is underlined in several studies conducted with laboratory and micro-Raman spectrometers including the micro-Raman analysis of emeralds [20], differentiation of corundum rubies and forgeries [21], etc. Casadio *et al* in 2007 [22] used several analytical techniques, including micro- and mobile Raman spectroscopy, to study Chinese jades. Barone *et al* in 2014 [23] compared the micro- and mobile Raman approach on blue-ranged gems. Handheld x-ray fluorescence was also used in this study. Mobile Raman instruments were employed for research on gemstones of the sceptre of the Faculty of Science, Charles University (Prague, Czech Republic) [24]; on gemstones on a Torah shield [25] and on the Loreto treasury objects and on the Ring Monstrance (Prague, Czech Republic) [26]; and on jewels from the Messina Regional Museum [27] and from the Paolo Orsi Regional Museum (Siracusa, Sicily) [28].

A valuable tool for geological analysis using Raman spectroscopy is the mapping procedure. When Raman mapping an object, the chemical information can be associated with its spatial distribution. Thus, spatial differences (if any) can be visualised with respect to the molecular synthesis of the material under study. Raman mapping was used for investigating coesite (SiO_2) inclusions in garnet of the Kokchetav Massif in Northern Kazakhstan [29] and (line-scanning) for visualising inclusion in garnets found in Eclogitic marbles, Alpine Corsica [30].

2.3 Selected topics of Raman spectroscopy for gemological purposes, including forgeries

2.3.1 Garnets

Among the gems with great historical importance, garnets occupy a prominent place. They were used as gems from antiquity, but they found their maximum relevance in the Middle Age, especially in the Longobard period in Western Europe. The name 'garnet' refers to a mineral group formed by silicates with a general formula $X_3^{2+}Y_2^{3+}(SiO_4)_3$. The most common terms are usually divided in two families: calcic garnets 'ugrandite' where X = Ca and Y = Cr (uvarovite), Al (grossular), Fe^{3+} (andradite) and aluminate garnets 'pyralspite' where Y = Al and X = Mg (pyrope), Fe^{2+} (almandine), Mn (spessartine). In nature they are usually found as solid-state solutions of different pure terms (end-members), mostly of the same family. The pure terms are very rare. Aluminate garnets have been largely used in jewellery thanks to their typical intense red colour. The colour is mostly due to the presence of Fe^{2+}. Even if it is possible to find beautiful red pyrope or spessartine garnets (i.e. garnets with pyrope or spessartine as main components), their red hue is due to the presence of the almandine term in the composition. A pure pyrope garnet is colourless, while pure spessartine is orange-yellow. Calcic garnets have been less used in jewellery due to their less attractive colour, ranging from orange-red to yellow-green in the grossular–andradite series. Some garnets of the 'hessonite' variety sometimes have a definite red colour and can be used in substitution of almandines. The only exception among the calcic garnets is the rare chromian term, uvarovite, possessing a very intense green colour, used to create beautiful gems. In order to correctly evaluate a garnet, to obtain information on its nature, provenance, and history, it is fundamental to determine its composition, which can be expressed as the amount of the different end-members. The traditional analytical techniques employed to obtain the chemical composition of garnets are usually destructive and time-consuming. Very often they require that the gem should be loose (not mounted on a jewel or artwork) and moved to the analysis laboratory.

Raman spectroscopy has been proposed as a routine tool for a quick, non-destructive analysis of garnets [31, 32] in order to obtain their molar composition starting from their spectra. The method is based on the assumption that the vibrations in garnets have a so-called one-mode behaviour. This means that in the Raman spectrum of a garnet, which is a solid solution of different end-members, the wavenumber ν of each Raman peak is obtained as the linear combinations of the wavenumbers shown by such peak in the end-members (ν_{Alm}, ν_{Spe}, ν_{Pyr}, ν_{Gro}, ν_{And}, ν_{Uva}), weighted by their molar fraction x:

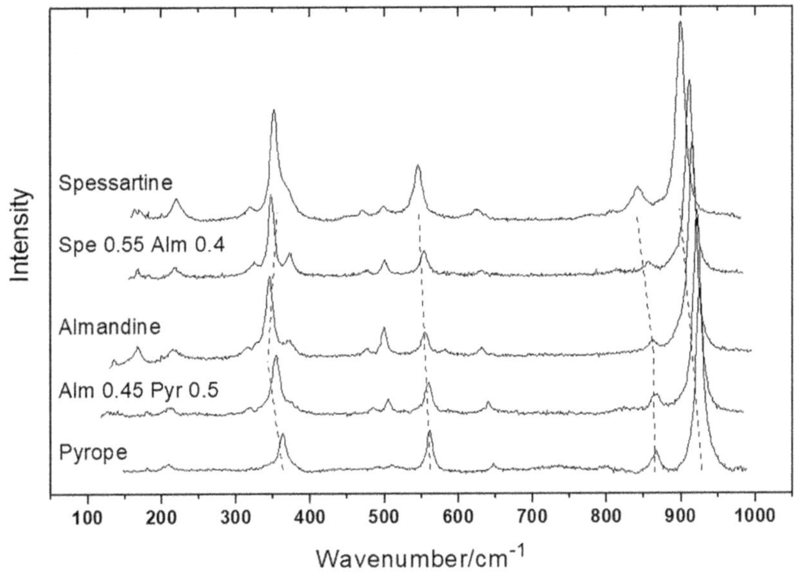

Figure 2.1. Raman spectra of aluminate garnets in the series pyrope–almandine and almandine–spessartine. Dashed lines are drawn as guide for eyes in order to show the shift of some peaks [31].

$$\nu = x_{\text{Alm}}\nu_{\text{Alm}} + x_{\text{Spe}}\nu_{\text{Spe}} + x_{\text{Pyr}}\nu_{\text{Pyr}} + x_{\text{Gro}}\nu_{\text{Gro}} + x_{\text{And}}\nu_{\text{And}} + x_{\text{Uva}}\nu_{\text{Uva}}$$

In figure 2.1 an example of the behaviour of the Raman peaks changing the composition from pyrope to almandine and from almandine to spessartine is shown.

The position of the Raman peaks of the most relevant end-members is well known from literature [31, 33–36]. After the collection of the Raman spectrum on a garnet, measuring the wavenumbers of a selected group of Raman peaks, it is possible to estimate the composition of the analysed garnet. In order to automatise the process, a Matlab® routine named 'Miragem' was prepared. Through this software it is possible to obtain the estimated chemical composition of the garnet giving as input the wavenumbers of four to six Raman peaks. An updated version of the software is freely available on request from the authors of the present chapter. The method was demonstrated to be able to obtain the composition of the garnets with a discrepancy (with respect to a standard elementary analysis such as by a wavelength dispersive x-ray spectroscopy (WDS) electron microprobe) usually less than 10% for the main component (figure 2.2) [31]. Worse results are obtained for uvarovite-rich garnets because vibrations involving Cr^{3+} have two-mode behaviour. Of course, the precision is lower than that obtained with an electron microprobe, but the composition is obtained in a very quick way (usually in a few seconds) and with a contactless, non-invasive technique.

The results presented in the early works [31, 32, 37–39] were obtained using high-performance laboratory micro-Raman spectrometers. In recent works, we observed that, despite their worse spectral resolution, the wavenumber accuracy obtainable with mobile spectrometers (after a suitable calibration) is enough to calculate the

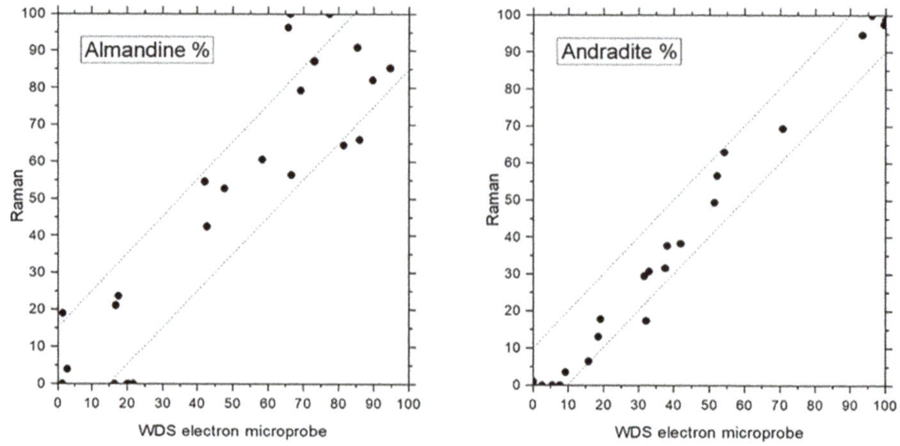

Figure 2.2. Comparison from the amount % of almandine in aluminate garnets and of andradite in calcic garnets, calculated from Raman data and measured by a WDS electron microprobe. The lines of ±15% for almandine and ±10% for andradite are shown. Adapted from [31] with permission from Elsevier.

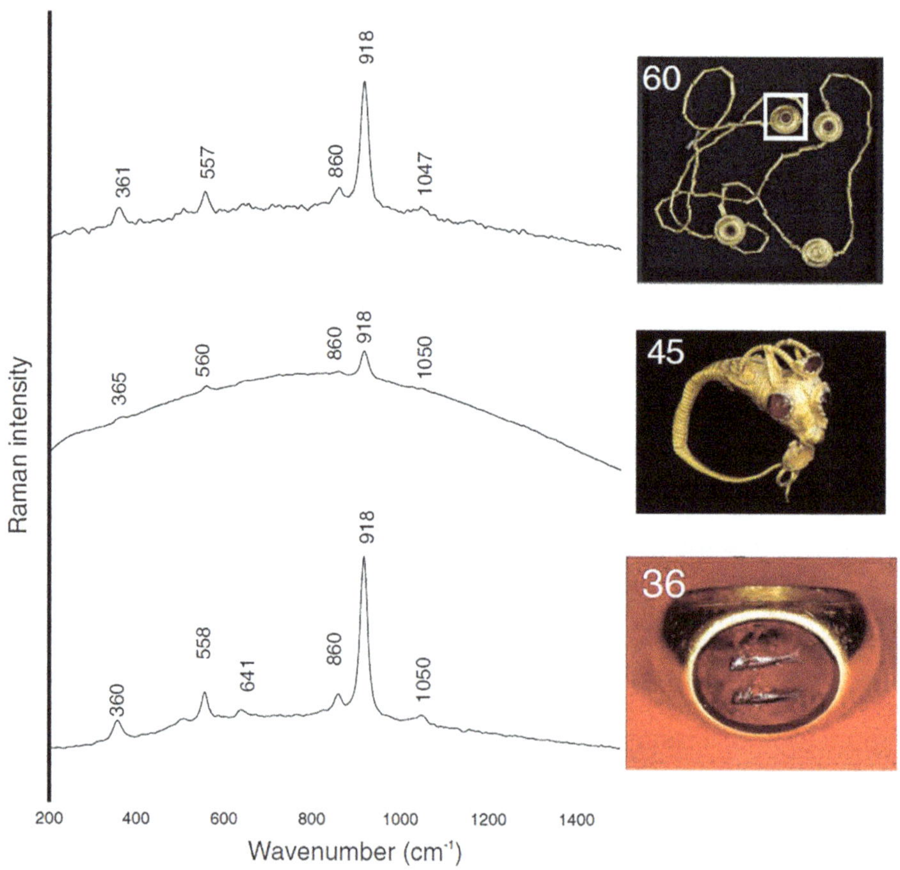

Figure 2.3. Raman spectra taken using a handheld Raman spectrometer operating at 785 nm of garnets mounted on precious Hellenistic jewels preserved in the Paolo Orsi Regional Museum (Siracusa, Sicily). Adapted from [28], copyright 2009 with permission from Elsevier.

composition of garnet. This possibility opens a wide range of applications in the field of gemology, in particular historical and archaeological gemology. It is now possible to make a quick determination of the composition of garnet gems directly in museums or even of gems mounted in complex objects that cannot be moved to a laboratory due to their value or conservation issues. As an example, during a short campaign of measurements performed to identify gems mounted on jewels from Magna Grecia preserved in the Paolo Orsi Regional Museum in Siracusa (Sicilia, Italy) [28], some garnets were identified by means of handheld Raman spectrometers. In figure 2.3 the spectra obtained with a DeltaNU spectrometer (with excitation at 785 nm) are shown. The typical spectrum of pyralspite garnets is visible. The quality of the spectrum was enough to use the Miragem routine to identify all the gems' compositions as prevalently almandine (70%–80%), with spessartine as the second member.

Another important investigation was performed on a Longobard treasure found in Lodi Vecchio (Lombardia, Italy): many golden jewels containing garnets (different rings and a beautiful necklace cloisonné) were analysed (figure 2.4). The spectra were obtained with a EnSpectrRaport handheld spectrometer operating at 532 nm. Even in this case the obtained spectra were of high quality. All the measured garnets showed nearly identical Raman spectra with very close peak positions (figure 2.4). The Miragem software gave for all the analysed garnets almandine as the prevalent term (from 64% to 76%), but in this case the second term was pyrope. Taking into

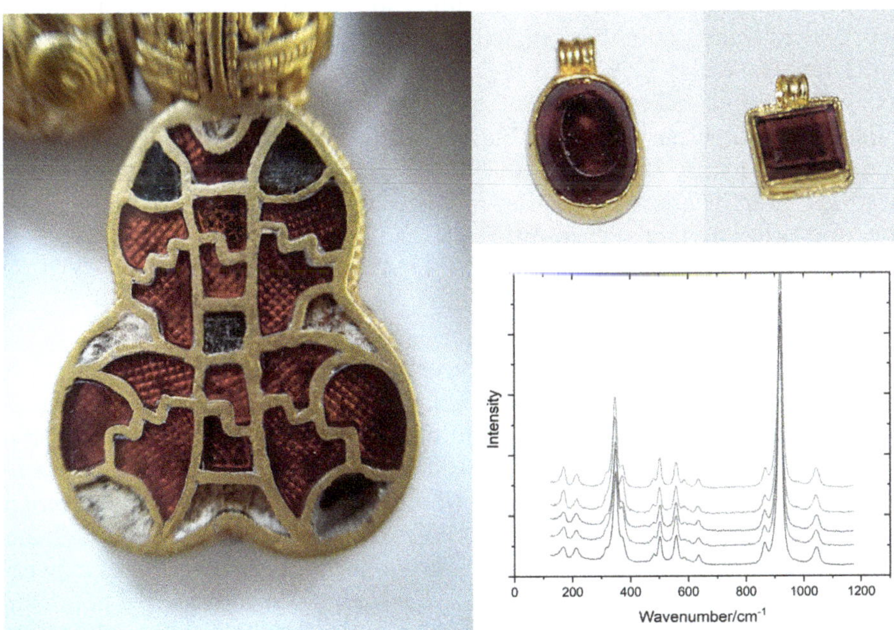

Figure 2.4. Analysed Longobard jewels coming from Lodi Vecchio (Lodi, Italy) containing almandine garnets and Raman spectra of some of the garnets. Reproduced by permission of Ministero della Cultura – Soprintendenza Archeologia Belle Arti e Paesaggio per le Province di Cremona Lodi e Mantova, Italy.

Figure 2.5. One of the authors (P.V.) performing Raman analysis on the garnets present on a pegmatite outcrop in Valchiavenna (Lombardia, Italy) using a EnWave EZRAMAN-DUAL spectrometer.

account the uncertainty of this method (around 10%) and the natural compositional variation of garnets, we could suggest a common origin for all the analysed gems.

With the use of portable spectrometers, it is also possible to study garnet gemstones directly on outcrops for a compositional comparison with cut gems. A test study of pegmatitic outcrops containing almandine garnets was made in the Alps of Valchiavenna (Lombardia, Italy) using a fibre-optics double laser (785 nm and 532 nm) EnWave EZRAMAN-DUAL, transported on a backpack through rivers, woods, and valleys (figure 2.5). Even in this case, the composition of prevalent almandine garnets was obtained. Although the instrument was heavier than a handheld one, the ability to choose between two different laser sources allowed the identification of a large number of other minerals (beryls, titanites, quartz, and other minerals) even in the presence of strong fluorescence.

2.3.2 Jade

Raman spectroscopy can be successfully used even for the identification of semi-precious stones such as jade. Jade is a hard translucent green rock used from prehistoric times to create jewels, sculptures, and art objects; it is one of the most important materials in different cultures around the world, from South America to the Far East [40]. Different rocks, composed of different silicate minerals, are recognised as jade. The most precious and important kinds, often considered 'true' jades, are the jadeitic jade and the nephritic jade. The first one is made by jadeite, a pyroxene mineral with the formula $Na(Al,Fe^{3+})Si_2O_6$; the second one is composed of amphibole minerals belonging to the so-called nephritic series $Ca_2(Mg_xFe^{2+}{}_{1-x})_5Si_8O_{22}(OH)_2$. Tremolite $Ca_2Mg_5Si_8O_{22}(OH)_2$ is the magnesium end-member of that series, while ferro-actinolite $Ca_2Fe^{2+}{}_5Si_8O_{22}(OH)_2$ is the iron-rich end-member. Other hard green rocks have been sometimes considered to be less precious varieties of jade, such as the omphacite (omphacitic jade), very close to the jadeitic jade but in large part composed of pyroxenes different from jadeite (kunzite, hiddenite, augite, diopside, and others), or even rocks made by serpentine minerals ('serpentine jade' or 'new jade').

Raman spectroscopy is able to distinguish very easily among pyroxenes [41–43] including jadeite [44], amphiboles [45–48], serpentines [49, 50], and other green silicates, even using portable equipment (figure 2.6).

Using Raman spectroscopy, it is also possible to identify the different kind of pyroxenes [41–44], helping in the identification and discrimination of the different kinds (and value) of the jadeitic-omphacitic jades.

In the case of the precious nephritic jade, in order to have some clues to its value, history, and provenance it is fundamental to know its chemical composition. The nephritic series tremolite–ferro-actinolite has a general formula $Ca_2(Mg_xFe^{2+}{}_{1-x})_5Si_8O_{22}(OH)_2$. To identify the different terms in the series we can use the X ratio defined as $X = Mg/(Mg+Fe^{2+})$. For $x>0.9$ we have the mineral tremolite, in the range $0.9>x>0.5$ we have actinolite, and for $x<0.5$ we have ferro-actinolite. Usually, terms between tremolite and actinolite are present in precious jades. The chemical composition can be obtained in a quick and completely non-invasive way by using Raman spectroscopy [45]. We can obtain information on the X ratio by analysing both the low-wavenumber (fingerprint region) and the high-wavenumber (OH-stretching region) parts of the spectrum. The fingerprint region of the spectrum (up to 1100 cm^{-1}) is characterised by a strong Raman peak at nearly 675 cm^{-1}, attributed to Si–O–Si symmetrical stretching with A_g symmetry (figures 2.6(a)–(c)).

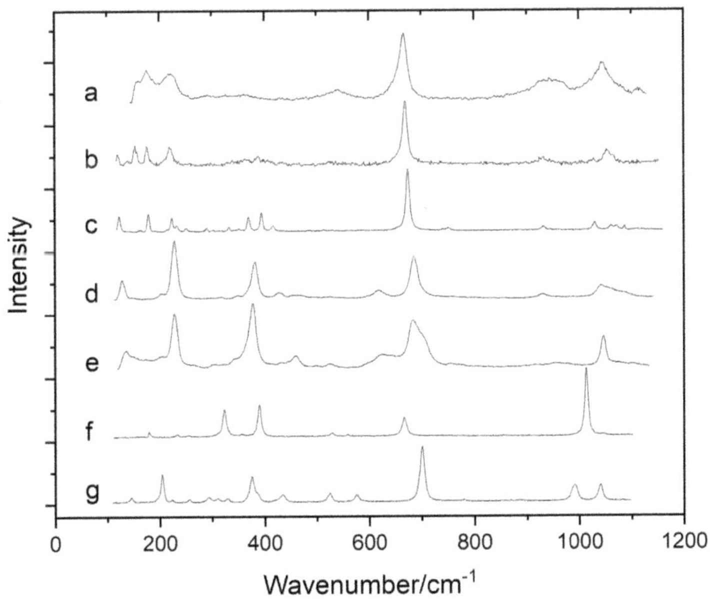

Figure 2.6. Raman spectra of some silicates that can be present in jades. Amphiboles belonging to the nephritic series: (a) ferro-actinolite $X = 0.42$, (b) actinolite $X = 0.68$, (c) tremolite–serpentine spectra. Minerals of the serpentine group: (d) lizardite, (e) antigorite. Pyroxenes: (f) diopside, (g) jadeite. The jadeite spectrum was taken from the RRUFF database (no. 050220).

Substituting lighter Mg^{2+} ions with heavier Fe^{2+} ions, the peak position downshifts from 675 cm^{-1} in pure tremolite to 667 cm^{-1} at the actinolite–ferro-actinolite limit. The shift of that peak is linear with the value of X: this allows us to obtain the chemical composition of a nephritic jade by the formula

$$X = 0.066x(\nu - 659.3)$$

where ν is the wavenumber in cm^{-1} of the main Raman peak in the fingerprint region [45]. The same information can be obtained by the analysis of the high-wavenumber region of the spectrum (3400–3800 cm^{-1}) where one to four sharp peaks are present, depending on the X value, attributed to the stretching vibrations of the OH groups. The wavenumber of the OH groups depends on the nature of the ions surrounding the OH groups. In the structure of the amphiboles belonging to the nephritic series, the OH groups are located in a site near to three sites (called C-sites) occupied by divalent metals [45–48]. When all the C-sites are occupied by Mg^{2+} (Mg–Mg–Mg combination) as in pure tremolite ($X = 1$), all the OH groups vibrate at the same frequency, and only a Raman peak is obtained, at 3675 cm^{-1}. As the amount of Fe^{2+} increases (X decreases), moving from tremolite to actinolite, the population of the C-sites changes, and some OH groups will vibrate at different frequencies, according to different combinations of Fe^{2+} and Mg^{2+}. Three new Raman peaks then arise at lower wavenumbers, attributed to the Mg–Mg–Fe, Mg–Fe–Fe, and Fe–Fe–Fe combinations. Some examples are shown in figure 2.7. The intensity of these Raman bands depends on the relative amounts of

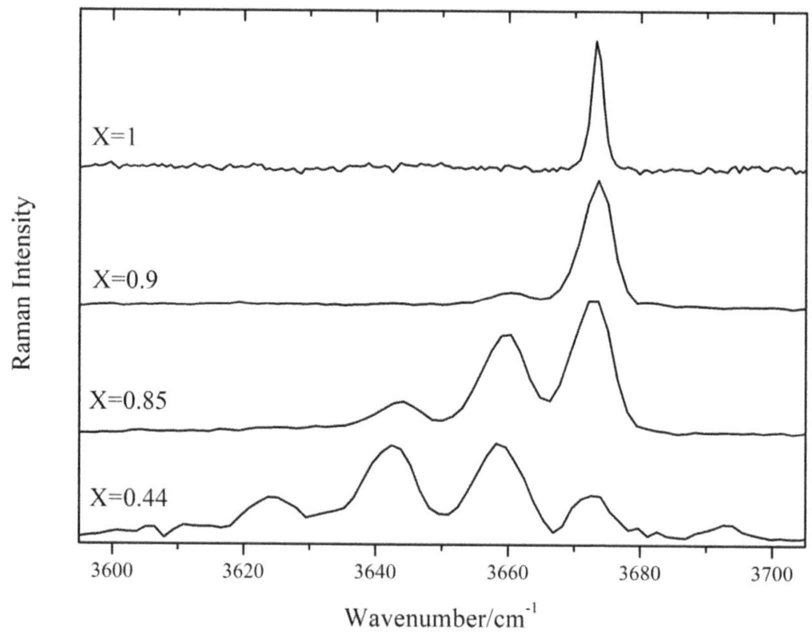

Figure 2.7. Raman spectra of the OH-stretching region of members of the tremolite–ferro-actinolite series at different X values. Reproduced from [45], CC BY 4.0.

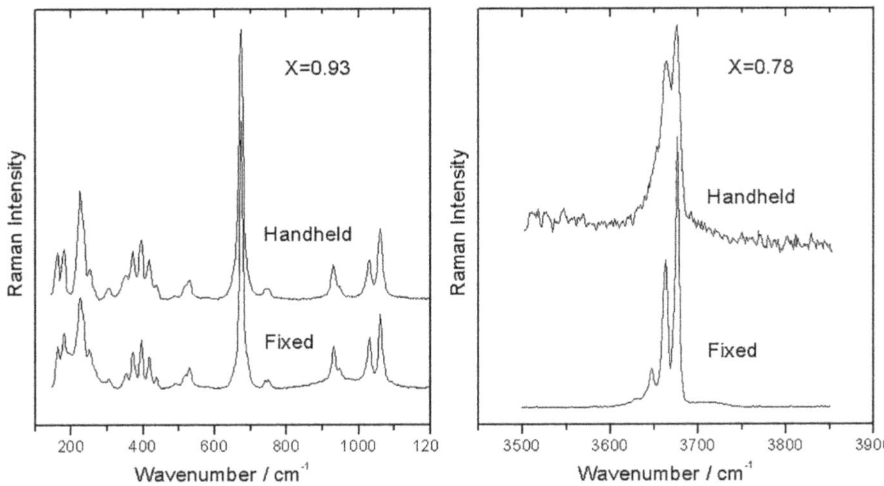

Figure 2.8. Comparison of Raman spectra in the low-wavenumber and in the OH-stretching region taken with fixed and mobile spectrometers. Reproduced from [45], CC BY 4.0.

Fe^{2+} and Mg^{2+} and can be used to determine the X value. A simple relation, based on a statistical model, using the ratio of the areas of the first two bands, was proposed [45]:

$$X = (A_{12})/(1/3 + A_{12})$$

where A_{12} is the ratio between the area of the band at 3675 cm^{-1} and the one at 3660 cm^{-1}.

Both methods have been shown to be reliable and usable also with portable spectrometers, even if some limitation due to their lower spectral resolution and limitation in the spectral range should be taken into account. In figure 2.8 a comparison between spectra obtained with fixed and mobile spectrometers in both spectral regions is shown.

The majority of mobile Raman spectrometers do not allow the analysis of the OH-stretching region, so the method based on the areas of the OH bands can be used only when using wide-range spectrometers. On the other hand, the obtaining of X from the wavenumber of the main Raman peak in the fingerprint region requires a good spectral accuracy and a perfect wavenumber calibration, not always easy with mobile spectrometers. So, the best method to obtain the nephritic jade composition using mobile instrument depends on the technical characteristics of the spectrometer used.

2.3.3 Beryls

Beryl is a cyclosilicate with formula $Be_3Al_2Si_6O_{18}$. Its chromatic varieties are among the most valuable and appreciated coloured gems. Emerald is the deep green variety, whose colour is due to the presence of Cr^{3+} impurities. Aquamarine is the blue one; in this case the colour is due to a combination of V and Fe ions. Other valuable coloured beryls are the pink morganite and the yellow heliodor; the colourless goshenite is not often used as jewel. When analysing a beryl from a mineralogical, gemological, or archaeological point of view, it is important not only to identify the mineral species, a very easy task with Raman spectroscopy, but also to have indications of its origin. The most important indications are discrimination between synthetic and natural beryls and clues to the beryl's possible provenance.

Figure 2.9 shows the typical Raman spectrum of a beryl [20].

The fingerprint region (50–1100 cm^{-1}) is very characteristic, with a strong peak at 682 cm^{-1}. Two important peaks are present at 1009 cm^{-1} and 1066 cm^{-1}. The last one, which can be attributed to Be–O or Si–O stretching [51, 52], is very sensitive to the presence of alkali ions in the structural channels of beryls [53, 54]. In beryls grown in alkali-rich rocks (such as the schists) this peak is broader (more than 18 cm^{-1}) and at higher wavenumbers (1069–1073 cm^{-1}). In beryls with low alkali content (such as the ones grown in granitic pegmatites) the peak width is lower and is found at wavenumbers between 1068 cm^{-1} and 1070 cm^{-1}. Hydrothermally grown synthetic emeralds are alkali-free and this peak is very sharp (11–14 cm^{-1}) and at even lower wavenumbers (1067–1068 cm^{-1}) [20].

A further, more defined, indication of the origin of beryls is obtained from the analysis of the OH-stretching region of the Raman spectrum. Beryl does not have water in its chemical formula, but usually it hosts water molecules in the channels present in its structure, formed by the stacking of six-membered rings of SiO_4tetrahedra. In figure 2.10 the spectra of the OH-stretching region of different beryls are shown. When no alkali

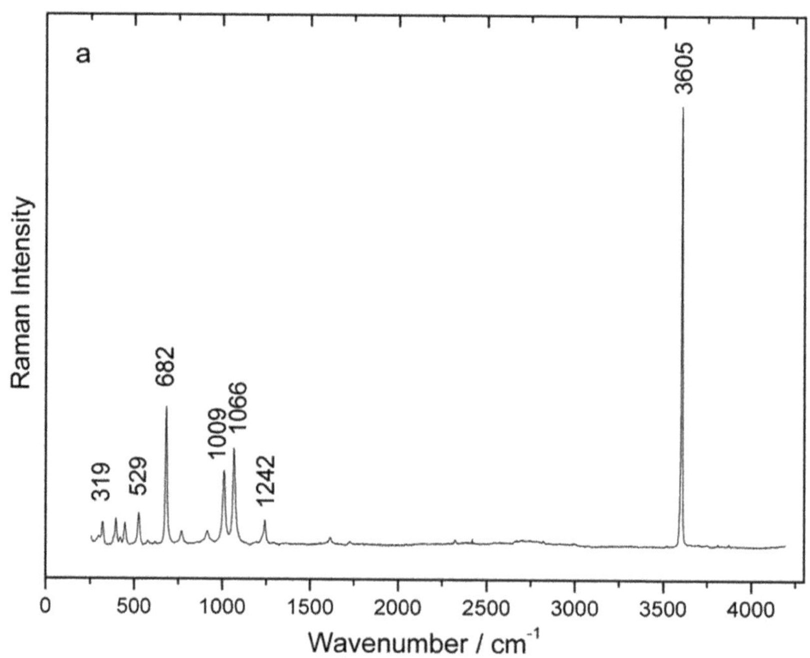

Figure 2.9. Raman spectrum of a beryl. Reproduced from [20] by permission from John Wiley & Sons, copyright 2014.

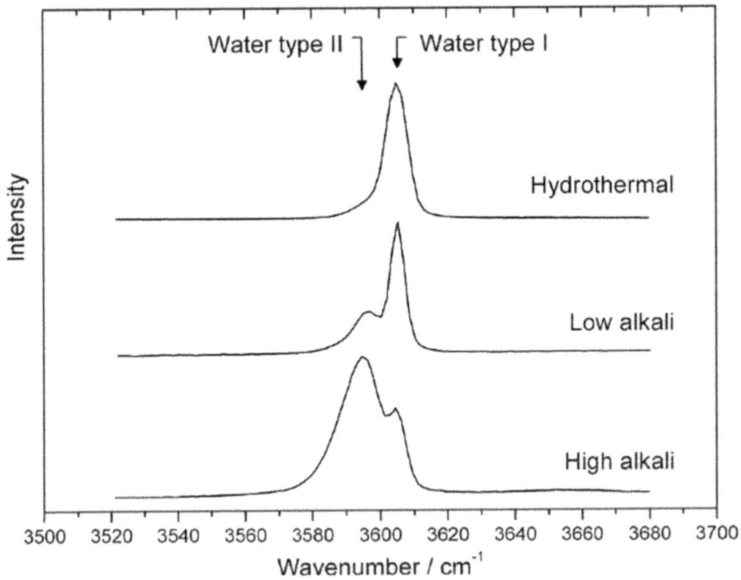

Figure 2.10. OH-stretching region of the Raman spectrum of different beryls. Reproduced from [20] by permission of John Wiley & Sons, copyright 2014.

ions are present, the water molecules are oriented with their symmetry axis perpendicular to the channel ('water type I'). When alkali ions are present in the channel, the neighbouring water molecules are oriented with the oxygen pointing to them ('water type II'). Due to the interaction with the alkali ions, water molecules of type II vibrate at lower frequency than water of type I. For that reason, in the OH-stretching region the Raman spectrum of beryls shows up to two different Raman peaks: one slightly over (water type I) and one slightly under 3600 cm^{-1} (water type II) (figure 2.10). In 'high-alkali' beryls coming from schists rocks the type II peak prevails, the contrary for the 'low-alkali' beryls. In synthetic beryls obtained by hydrothermal methods, only the type I peak is present. In synthetic emeralds obtained by melt, no water is present and no OH-stretching peaks are observed [20, 53, 54].

Mobile Raman spectrometers, both handheld and portable ones, with or without fibre optics, have demonstrated their capacity to analyse beryls directly on outcrops [16]. An extensive comparison between different models was performed in order to evaluate advantages and disadvantages of different configuration. When analysing beryls with Raman spectroscopy it is important to take into account the presence of fluorescence, which can be very high due to the presence of Cr^{3+} ions. That means that fluorescence will be very strong in emeralds (especially in some synthetic ones, very rich in Cr^{3+} chromophores), but also in some aquamarines. Cr^{3+} ions produce the maximum fluorescence output in a doublet between 680 nm and 684 nm. For that reason, the Raman spectra produced by exciting with a laser diode at 785 nm (the most diffuse solution in mobile spectrometers) are greatly affected. In this particular case, the use of a doubled Nd:YAG laser at 532 nm (the other standard choice for mobile spectrometers) is better, being less affected by the Cr^{3+} fluorescence. A handheld

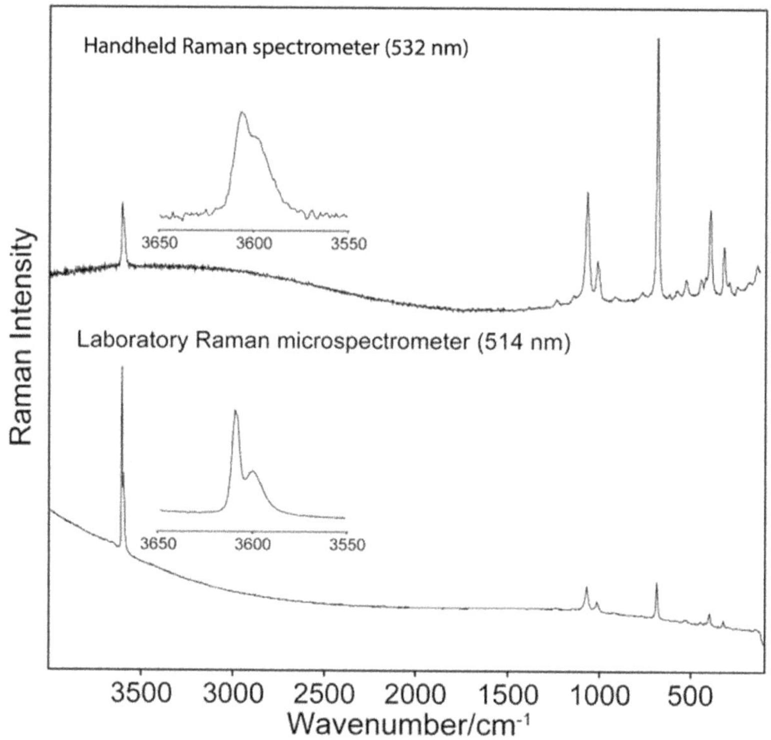

Figure 2.11. Comparison between the Raman spectra of an aquamarine beryl taken with a fixed and a mobile spectrometer. Reproduced from [16], with permission from John Wiley & Sons, copyright 2017.

spectrometer working in the near IR, using the principle of the sequentially shifted excitation, completely removes the fluorescence effect [16]. When working with an instrument whose spectral range includes the OH-stretching region (not very common among the portable ones) it is possible to study the presence of the alkali ions, and then the origin of the beryls, even *in situ*, using mobile spectrometer. The bands of type I and type II water overlap, but the spectral profile still allows us to discriminate between alkali-rich and alkali-poor beryls. In figure 2.11 a comparison between spectra obtained with mobile and fixed spectrometers is shown.

The use of Raman spectroscopy allows us to discriminate between beryls and similar gems and in particular allows the identification of the rare Cs-bearing mineral pezzottaite ($Cs(Be_2Li)Al_2Si_6O_{18}$), discovered in 2002, presenting strong similarities in aspect, chemical composition, and structure to beryl [55].

2.3.4 Corundum and other gemstones

Corundum is the aluminium oxide α-Al_2O_3 which is very well represented by its most important chromatic varieties sapphire (blue) and ruby (red) in the gemological world. Cr^{3+} impurities are the origin of the red colour in ruby, while an interplay between iron and titanium ions generates the blue colour of sapphire.

A mobile Raman spectrometer can be used for a quick identification of corundum gems, discriminating them from their simulants [21, 23], even in mounted gems or in museums [27]. It should be taken into account that the strong fluorescence of Cr^{3+} ions, present especially in rubies but always present in corundum, can produce a very high background signal in the red and near-IR part of the spectra, masking the Raman peaks. In that case, sometimes a Raman spectrum with just one or two characteristic peaks emerging is obtained. Particular attention should be paid in this case to the discrimination of ruby and one of its historical simulants, spinel, because the main peaks appear at similar positions: 417 cm^{-1} and 645 cm^{-1} for ruby and 409 cm^{-1} and 664 cm^{-1} for spinel [27]. As in the case of emeralds, in order to reduce the Cr^{3+} fluorescence, the use of instruments equipped with short-wavelength lasers can be helpful.

The use of micro-Raman instruments, equipped with confocal microscopes, allows the study of micrometric solid and fluid inclusions. In addition to the possibility of identifying synthetic corundum and detecting traces of enhancement treatments, as discussed later, the analysis of inclusion can be very helpful in the identification of the origin of the gem. As an example, in a cut ruby a peculiar group of crystalline inclusions was found, composed of rutile TiO_2, zircon $ZrSiO_4$, and baddeleyite ZrO_2, all well identified by their characteristic Raman spectra (figure 2.12) [21]. In addition, the Raman spectra of baddeleyite inclusions revealed

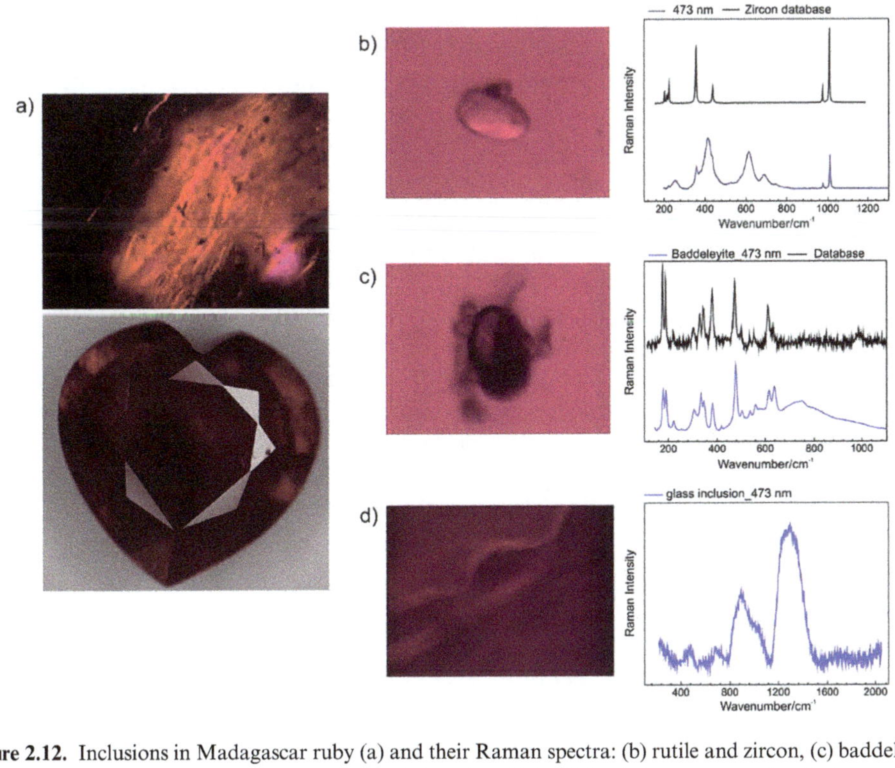

Figure 2.12. Inclusions in Madagascar ruby (a) and their Raman spectra: (b) rutile and zircon, (c) baddeleyite, (d) glass filler. Reproduced from [21], with permission from John Wiley & Sons, copyright 2016..

the sharp fluorescence peaks of Dy, Er, and Eu impurities. This combination of inclusions and rare earth elements is typical of heated rubies from Madagascar [56].

Recent works have shown the effectiveness of mobile Raman spectrometers, with different configurations, in the detection of many different mounted and loose cut gems and gemstones, such as tanzanite (the blue-violet variety of zoisite) [19]. The use of a portable Raman spectrometer working in the near-IR range with the technique of sequentially shifted excitation allowed the obtaining of representative Raman spectra of many highly fluorescent minerals used as gemstones, including chrysoberyl, taaffeite, and musgravite [17].

2.3.5 Raman and photoluminescence emission

Raman spectroscopy is an ideal tool for the analysis of gemstones, including pearls and corals. This non-destructive technique can give structural information about the gemstones under study. Moreover, when shifting between excitation sources for analysis, useful details of the nature of the gemstone can be revealed.

To complement Raman spectroscopy, gemologists are using another approach to accompany their analysis. Raman and photoluminescence spectra can be collected with the same Raman spectrometer and give information about the formation of the gemstone. In order to differentiate the actual Raman vibrations from the fluorescence bands, one should be aware of the physical phenomena occurring between Raman scattering and fluorescence emission.

Raman scattering, or inelastic scattering of light, is a weak phenomenon occurring (among other interactions) when a molecule interacts with the electromagnetic radiation thus one can describe it as a molecule–photon interaction [57]. Visualised in a basic energy diagram, the Raman scattering is associated with processes and steps between the ground/vibrational states and the virtual states. In practice, the Raman band positions are independent of the excitation wavelength. What is really changing with changing the wavelength of analysis is the fluorescence emission. With sufficient excitation energy, a molecule can be excited to a higher electronic energy state, and after relaxation, end up again in the ground state. This is true when Raman scattering is depicted in Raman wavenumbers in cm^{-1} (energy difference) at the x-axis of a spectrum. In a photoluminescence spectrum the processes are translated in wavelength in nm at the x-axis; the narrow fluorescence bands do not shift for different excitation lasers, but the Raman bands do.

Indeed, photoluminescence spectra have proven useful in various studies, especially when combined with classical Raman spectroscopy. Bersani *et al* in 2014 [20] used the 632.8 nm line of a micro-Raman system to study photoluminescence bands of Cr^{3+} in emeralds. Karampelas *et al* in 2021 [58] used a 633 nm laser to investigate the narrow bands of Cr^{+3} in emeralds/beryls and the broad feature related to chromium centres of gems found on a pendant from Einsiedeln Abbey in Switzerland. Coccato *et al* in 2021 [19] employed a 473 nm, a 532 nm, and a 785 nm laser excitation (using benchtop and portable instruments) to identify and differentiate photoluminescence bands in Raman spectra linked to rare earth elements, Cr^{3+}, etc when studying zoisite and tanzanite. Eaton-Magaña *et al* in 2021 [59] discussed the possibilities of Raman and photoluminescence mapping for gemological purposes.

2.3.6 Glass

Raman spectroscopy has proven an extremely valuable tool for the identification of glassy matrices. Generally, the Raman spectrum of a glassy material is influenced by the bending and stretching vibrations of the Si–O network [60]. In glass, SiO_4 tetrahedra are connected together with oxygens [61]. Additions may alter this network by introducing non-bridging oxygens in the SiO_4 tetrahedra connections [60–62].

The bending vibrations can be found at around 500 cm^{-1}, while the Si–O stretching ones can be observed at around 1000 cm^{-1}. The ratio of the two areas is described in the literature as the polymerization index ($I_P = A_{500}/A_{1000}$) and provides information on the glass type/family and fabrication temperature [60, 61].

Figure 2.13. Longobard drinking horn and necklace coming from Lodi Vecchio (Lodi, Italy). The spectra, taken on the blue glass of the horn and on the large green pearl of the necklace, are typical of soda-lime glasses. Reproduced by permission of Ministero della Cultura – Soprintendenza Archeologia Belle Arti e Paesaggio per le Province di Cremona Lodi e Mantova, Italy

In addition, the shape of the glass bands and the position of their maxima allows us to categorise them, discriminating among different production technologies and among the different glass modifiers employed [60, 63]. Moreover, from the deconvolution in five Gaussian bands of the Si–O stretching band it is possible to estimate the ratio between bridging and non-bridging oxygen atoms in the glass structure [60, 63].

As an example, figure 2.13 shows the spectra of the glasses composing two important archaeological Longobard jewels, found in Lodi Vecchio (Lodi, Italy) and preserved by the Soprintendenza Archeologia, Belle Arti e Paesaggio di Milano. They come from two different objects, a blue drinking horn and a green pearl of a glassy necklace, but their spectra are very similar, in terms of shape and ratio of the stretching and bending bands, indicating a very similar composition and structure. The polymerisation index is low, around 0.55 for both glasses, indicating a low fusion temperature. Both of them can be attributed to the family of soda-lime glasses, largely used in the Longobard period.

2.3.7 Pearls and corals

Pearls are fascinating organic gems, which have been used in jewellery since ancient times. Without needing to be shaped or carved, pearls were employed in the production of magnificent necklaces, pendants, brooches, rings, earrings, and other ornaments. Corals, belonging to the *Hydrozoa* and *Anthozoa* classes of marine invertebrates, have been used in the production of jewels, similarly to pearls. Besides jewellery, precious carvings and sculptures have been created with corals. The popularity of jewels made of pearls and corals is evidenced through the ages by archaeological evidence as well as by representations showing pearl and coral jewels—e.g. in portraits and illustrations—and historical testimonies [64–66].

Molluscs form pearls through biomineralisation processes, secreting calcium carbonates embedded in an organic matrix in concentric layers. Corals, as well as mollusc shells, are marine organisms which mineralise their own calcium carbonate structure. In pearls, when the mantle tissue is injured, e.g. from an animal attack, some cells of the external mantle are moved into the deeper layer, where they constitute a cyst or *pearl sac*. The production of calcium carbonate by the mantle cells—containing the so-called *nacre*, or *mother of pearl*—determines the accumulation of material here, originating the pearl. Pearls can be natural—naturally secreted by the molluscs from their mantle tissue—or freshwater/saltwater cultured—produced by the marine organisms after a human intervention consisting of an implantation of epithelial tissue in the mantle or in the gonad for stimulating the production of the pearl [67]. The composition of a pearl is 93%–95% of calcium carbonate, 5% of organic matrix, and 1% of moisture. The organic matrix is made of polysaccharides and proteins (*conchiolin*). Distinctive colours of pearls may range from white to intense tones of yellow, orange, pink, or purple. An example of natural and cultured freshwater and saltwater pearls is given in figure 2.14, showing pearls with different colours.

In pearl and coral analysis, both the biogenic carbonate structure and the pigments can be extensively investigated through Raman spectroscopy. As with

Figure 2.14. Picture of natural and cultured freshwater and saltwater pearls.

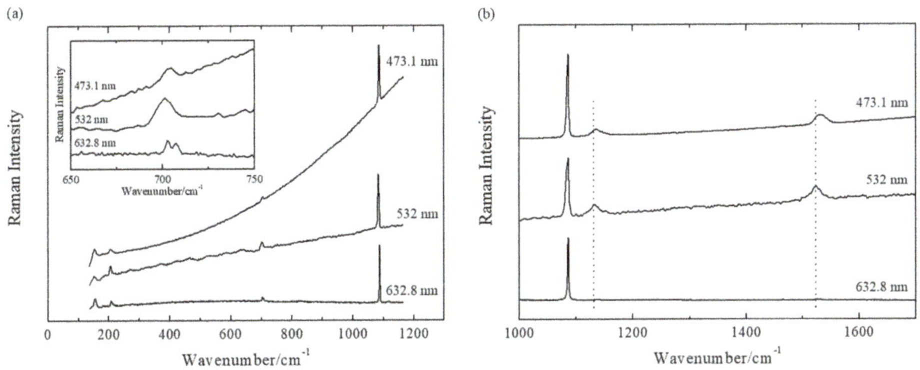

Figure 2.15. (a) Raman spectra collected on a white Japanese cultured saltwater pearl obtained at different excitation wavelengths: 632.8 nm, 532 nm, 473.1 nm. In the inset, the typical doublet of aragonite occurring at 702 cm^{-1} and 706 cm^{-1} is enlarged. (b) Raman spectra collected on a Chinese cultured freshwater pink pearl obtained at different excitation wavelengths: 632.8 nm, 532 nm, 473.1 nm. Adapted from [68].

other gemological objects, Raman analyses can be performed directly on the jewels with a non-destructive approach. In the structure of pearls, the three polymorphs of calcium carbonate may be distinguished by their characteristic Raman signals, and information on the orientation of the biogenic material may be obtained by means of polarised micro-Raman measurements. Nacre is composed of flat polygonal crystals of aragonite with their *c*-axes perpendicular to the layer plane [9].

Raman spectra collected on some pearls depicted in figure 2.14 evidence their aragonitic structure (figure 2.15(a)). The strongest peak occurs at 1086 cm^{-1}, corresponding to the stretching mode of the carbonate ion (CO_3^{2-}). The doublet occurring at 702 cm^{-1} and 706 cm^{-1} (inset of figure 2.15(a)), related to the bending mode of the CO_3^{2-} group, is typical of aragonite, which can be distinguished from the other calcium carbonate polymorphs, calcite and vaterite. It is noteworthy that, since the spectral resolution depends on the excitation wavelength, the doublet of aragonite can be better resolved using longer laser wavelengths (632.8 nm), whereas for the 532 nm and 473.1 nm excitations only a single band is obtained [68].

Interestingly, the aragonite peak at 855 cm^{-1}, corresponding to the out-of-plane bending of the CO_3^{2-} group, cannot be observed in natural aragonitic pearls. As shown in figure 2.16, in aragonite crystals this peak is detected only when the

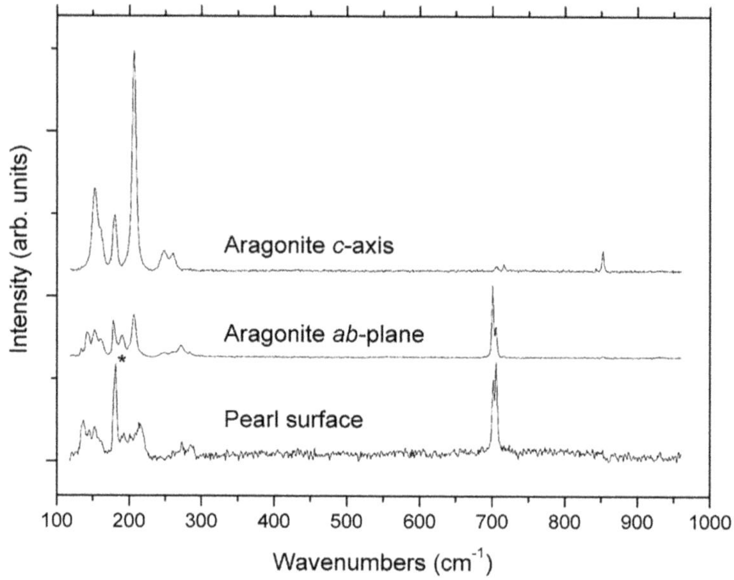

Figure 2.16. Raman spectra obtained on an aragonite crystal, with the laser polarisation parallel to the *c*-axis and to *ab*-plane, and on the surface of a pearl. The spectra are normalised on the strong 1087 cm^{-1} peak (not shown). The asterisk refers to the plasma line. Reproduced from [9], copyright 2010 by permission from Springer.

laser polarisation is parallel to the *c*-axis, whereas it is absent for measurements on the surface of natural pearls, since the *c*-axis has a radial direction. Otherwise, this feature may occur in an aragonite paste, whose crystals are expected to be randomly oriented [9].

Calcite and vaterite have been usually found in the inner part of pearls, together with aragonite. However, to investigate the spatial distribution of the different calcium carbonate polymorphs underneath the surface—e.g. by performing micro-Raman mappings—a sectioning of pearls is required. In addition, vaterite has been suggested as indicative of poor quality of a pearl, since it has been observed in correspondence with white lustre-lacking blemishes on the surface, by performing Raman analysis in a non-destructive way [69].

The lustre and colours of pearls are usually given by organic pigments consisting of polyenic chains. Polyenes (polyacetylenes) are polyunsaturated linear organic compounds containing sequences of double and single C–C bonds [70, 71]. In corals, attractive shades from pink to red are ascribed to both polyenes and carotenoids. Carotenoids are methylated polyenes with characteristic methyl groups substituted on the polyenic backbone [65, 70]. The organic pigments are easily detected with Raman spectroscopy, even for very low concentration. Raman spectral features of polyenic chains consist of two intense bands related to the C=C (ν_1) and C–C (ν_2) stretching modes, occurring between 1450–1680 cm^{-1} and 1070–1210 cm^{-1}, respectively. In addition, weaker bands are observed due to the CH=CH (ν_3) in-plane rocking mode and CH=CH (ν_4) out-of-plane wagging mode, between 1285–1315 cm^{-1} and 1000–1015 cm^{-1}, respectively. Raman spectral features are influenced by differences in the polyenic chains, including chemical factors—e.g. the substitution with methyl groups or the chemical environment—and electronic factors—e.g. the conjugation length of the polyenic backbone, isomerism, or conformation [72]. Therefore, Raman spectra of carotenoids consist of the same four main vibrational modes of polyenes, which occur in slightly different spectral ranges: 1480–1580 cm^{-1} (ν_1), 1100–1200 cm^{-1} (ν_2), 1260–1290 cm^{-1} (ν_3), and 990–1015 cm^{-1} (ν_4). The substitution with methyl groups in the polyenic backbone determines a decrease in the π-conjugation, and shifts towards higher wavenumbers are observed [72–74].

The position of the Raman peaks of polyenes can be correlated with the intrinsic colour of pearls, corals, and shells. The maximum absorption for the lowest energy optical transition of molecules containing a polyenic chain shows a linear relationship with inverse conjugation, i.e. the number of conjugated double bonds (N). Therefore, shorter conjugated polyenic chains absorb from the yellow-green to the blue-violet part of the visible range, inducing pink to orange colours, respectively. Conversely, longer chains give absorption maxima at higher wavelengths. In several studies, an empirical relationship between the N and the position of ν_1 and ν_2 has been suggested [70, 72, 75, 76], as reported by the following expressions [70]:

$$\nu_1 = 258.5/(N + 160) \times 10^3 \quad (cm^{-1})$$

$$\nu_2 = 192/(N + 160) \times 10^3 \, (cm^{-1})$$

In shells, pearls, and corals, N is typically found in the range 7–14. Correlating the N value with Raman features gives information about their intrinsic colours, whose pigments are not found free in nature.

In pearls, usually a single colour is formed by a combination of several polyenes of different lengths [77]. A distribution of effective conjugation lengths results in a distribution of optical transition energies, as the maximum optical absorption is a function of the length of the conjugated double-bonds polyene chain. The coexistence of non-equivalent chains may produce the resonance Raman scattering effect, tuning the excitation laser wavelength towards the energy of the $\pi \rightarrow \pi^*$ electronic transition. Shorter excitation wavelengths should enhance the Raman signals of shorter chains, since they are close to shorter UV–vis maximum absorption wavelengths and vice versa [70, 71].

The nature of pigments in pearls consisting of polyenes appears in figure 2.15(b), showing the C = C (ν_1) and C–C (ν_2) stretching modes of the polyenic chains. The resonance Raman effect is evidenced by using shorter excitations. The ν_1 and ν_2 band positions shift as the number of the double bonds of the polyene chain changes. Within a distribution of chains with different lengths, a selective Raman resonance effect is observed as shorter chains show higher vibrational frequencies due to the increase of the force constants [70–72].

In resonance conditions, the enhancement of the Raman intensity enables the observation of clear features in the Raman spectrum up to >4000 cm^{-1}. In the case of pearls as well as corals, overtones and combination modes are detected with distinct bands, even if the fundamental bands are related to a distribution of chains with different effective lengths (figures 2.17(a) and (b)). According to the four main vibrational modes, overtones and combination modes can be assigned [3].

As described above, the colour of pearls and corals is usually given by mixture of polyenes, and different colours correspond to different conjugation lengths. In addition to that, other colouring agents have been detected with Raman spectroscopy in pearls and corals. Tahitian pearls have natural purple-grey colours from porphyrins [78]. Furthermore, treatments are also used to artificially colour both pearls and corals. In pearls, traces of dyes may be accumulated on surface blemishes or near the drill hole, and their Raman signals may be easily acquired. However, dye identification is not straightforward (figure 2.17(c)). Freshwater pearls have been also black dyed by immersion into a silver nitrate solution, which is oxidised into silver oxide, and their Raman signals can be detected [79]. The features of amorphous carbon have also been observed in black dyed freshwater pearls, as shown in the Raman spectrum of the black freshwater pearl in figure 2.17(a).

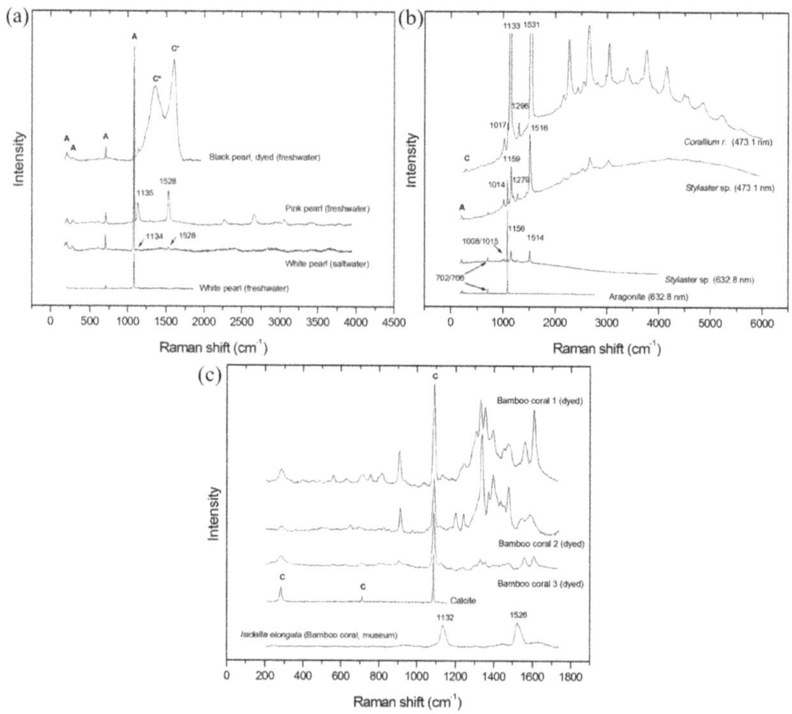

Figure 2.17. (a) Raman spectra collected on different freshwater and saltwater pearls, obtained at 473.1 nm. (b) Raman spectra of aragonitic coral *Stylaster* sp. and calcitic one *Corallium rubrum*. (c) Comparison between the Raman spectra acquired on some dyed commercial 'bamboo corals' and on natural *Isidella elongata* coral, obtained at 473.1 nm. Aragonite and calcite peaks are marked with A and C, respectively; amorphous carbon features are indicated with C*. Reproduced from [3] by permission of the publisher (Taylor & Francis Ltd, http://www.tandfonline.com).

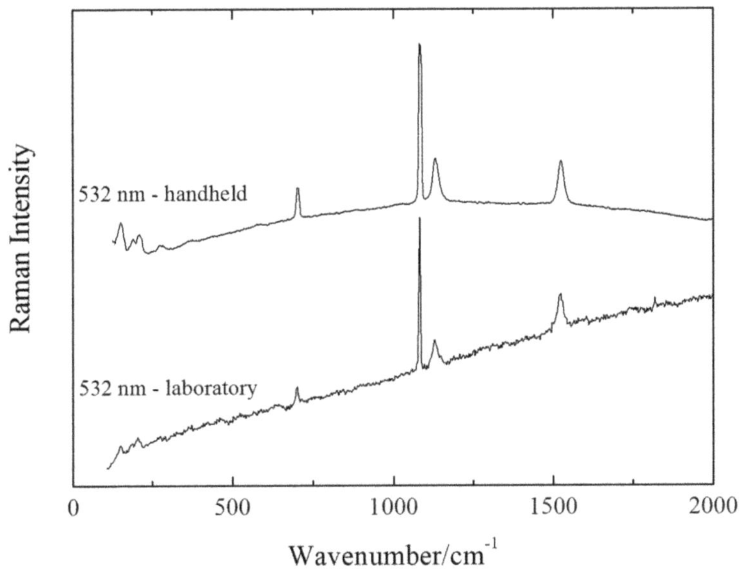

Figure 2.18. Comparison between Raman spectra collected with laboratory and handheld spectrometers working at 532 nm on a Chinese cultured freshwater pink pearl.

In pearls, the study of the structure and the colouring agents is also successfully achieved by using mobile spectrometers. Figure 2.18 illustrates the comparison between Raman spectra collected with mobile and laboratory instruments, proving the effectiveness of handheld devices in detecting the carbonate matrix features—i.e. aragonite peaks—together with the polyene bands—i.e. the ν_1 and ν_2 stretching modes of the polyenic chains. By using a 532 nm laser excitation wavelength, the resonance enhancement of the Raman signal is notably observed also with handheld instruments [68]. In addition, spectrometers covering spectral ranges up to ~4000 cm^{-1} may detect overtones and combination modes.

2.3.8 Forgeries

In gemology, the value of jewellery objects is strictly connected to their authenticity. Besides the intrinsic preciousness of the gem type, identifying forgeries is of paramount importance. Valuable gems have been tampered with, replaced with different materials or given treatments in order to magnify their aesthetic appearance. Simulants (also called imitations) are less valuable natural or synthetic materials used to substitute for a gem thanks to their similar aspect. Synthetic gemstones are artificial products, obtained with laboratory processes, with the same chemical composition and structure of the natural gems. Furthermore, treatments have been performed on gems to enhance their colours and their clarity: coating, filling, irradiating, and heating are some of the procedures used over time [66].

Regarding simulants, several types of materials have been used to imitate more valuable gems. Since ancient times, glasses of various colours have widely substituted for more precious emeralds, lapis lazuli, onyx, and turquoise. Glass processing technologies have been also refined with the development of lead glasses, which may simulate rubies and sapphires thanks to their higher refractive index [80, 81]. After the invention of celluloid, plastics-made jewels have been produced to imitate jades, corals, pearls, or even more precious emeralds, rubies, and diamonds. Plastics may be also dyed to give any desired appearance to the final objects. Diamond simulants are made of various materials: synthetic spinel, rutile, or garnet (e.g. YAG, yttrium–aluminium–garnet, and GGG, gadolinium–gallium–garnet), silicon carbide in the variety of moissanite, strontium titanate, or cubic zirconia—still the most widely used today. Among simulants, doublets and triplets consist of two and three different materials respectively, which have been glued together to simulate more valuable gems. In doublets, a colouring and a colourless materials are placed one on top of the other, or two materials are combined with a coloured glue. In triplets, the colouring layer is usually in between the two others [66, 79].

Synthetic gemstones are artificial products, obtained in the laboratory. The main crystal-growth techniques are melt-growth, solution growth, and vapor phase growth. In contrast to simulants, which imitate gems with different materials, synthetic gemstones have the same chemical structure as the natural ones and the same chemical composition, except for subtle differences at the level of impurities [82].

To improve gems' colour, treatments have been performed since ancient times [83]. In the past, the most widely used methods consisted of foiling with a metallic layer, coating with coloured lacquer or varnish, and dyeing by taking advantage of the porosity of the materials of some gems and filling with oils the porous or fissured surfaces of gems. Oils have been introduced in gems to hide surface imperfections—e.g. fissures, cracks—and to make the colours more brilliant. More recently, the use of oils has been replaced by filling with polymeric resins, which have a longer duration without altering their colours, as happens with oils [84]. Heat treatments are other processes for eliminating unpleasant shades or producing different colours in gems. Heat treatments may induce structural changes in the structure of gems or that of the inclusions embedded in them. The combination of heat and pressure treatments has been recently applied to produce synthetic gemstones (e.g. diamonds). Irradiation of gems is another procedure used for changing colours in gems [84]. Other enhancement treatments are not considered acceptable because they induce irreversible transformation. Not-allowed treatments have been obtained by partial melting of the surfaces of gemstones to hide the fractures. For melting, flux agents such as borax, calcium borate ($Ca_2B_2O_6{\cdot}5H_2O$), and calcium or sodium phosphate [$Ca_3(PO_4)_2$, Na_3PO_4] were used [85].

Raman spectroscopy is a powerful technique to investigate forgeries and discriminate between natural and synthetic gemstones. In recent decades, Raman spectroscopy has been applied in several gemological studies aimed at proving the authenticity of gems, suggesting that it can be considered a standard tool in gemological analysis. It is noteworthy that features of gems may be obtained both at the surface and even on the bulk underneath it with a non-destructive and non-invasive approach. Of course, Raman analyses may be performed also on mounted gems [16, 20, 21, 23, 80].

Micro-Raman spectroscopy proved to be effective in the study of inclusions—both fluid and solid—found in gems. Gases in bubble and channels can be identified by their Raman signals, such as the characteristic Fermi doublet at ~1290 cm^{-1} and ~1390 cm^{-1} of CO_2 or the peak at 1150 cm^{-1} of SO_2 [21, 23, 86]. A picture of an elongated CO_2 capillary found in a sapphire and the corresponding Raman spectrum are shown in figure 2.19. Moreover, solid inclusions have been found in gems, consisting of micrometric minerals (e.g. rutile needles, zircons crystals, baddeleyite crystals, etc) [21, 23, 79]. The identification of inclusions is of great importance in provenance and genesis studies—e.g. condition of formation, artificial treatments—since mineral inclusions are typical of specific places and rock matrices. In addition, characteristic inclusions are related to some synthetic processes (e.g. cryolite and silicon in rubies obtained with the Verneuil method) [87]. Among enhancement treatments, glass filling and heating can be investigated by micro-Raman analyses [21]. Another treatment that may be recognised with micro-Raman

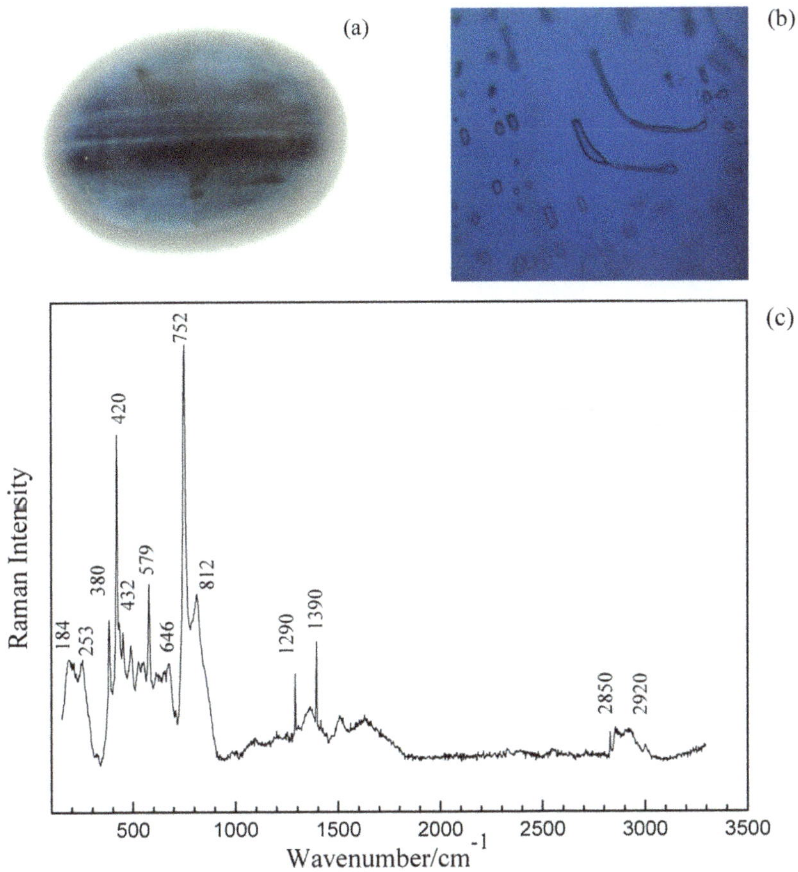

Figure 2.19. (a) Sapphire gem. (b) Micro-photo of the elongated capillary of CO_2 (field of view 60 μm × 60 μm). (c) Micro-Raman spectrum obtained at 473.1 nm on the sapphire, including the CO_2 and almond oil signals. Reproduced from [23] by permission from John Wiley & Sons, copyright 2014.

spectroscopy is filling with oils, usually applied to brighten gems by impregnating their fissures. Oils, as organic molecules, show characteristic C–H stretching bands in the spectral range 2800–3000 cm^{-1}, and filler oils used in jewellery may be distinguished and identified. In figure 2.19(c), the Raman signals of almond oil have been found in sapphire [23]. Other organic materials used as fillers, such as waxes,

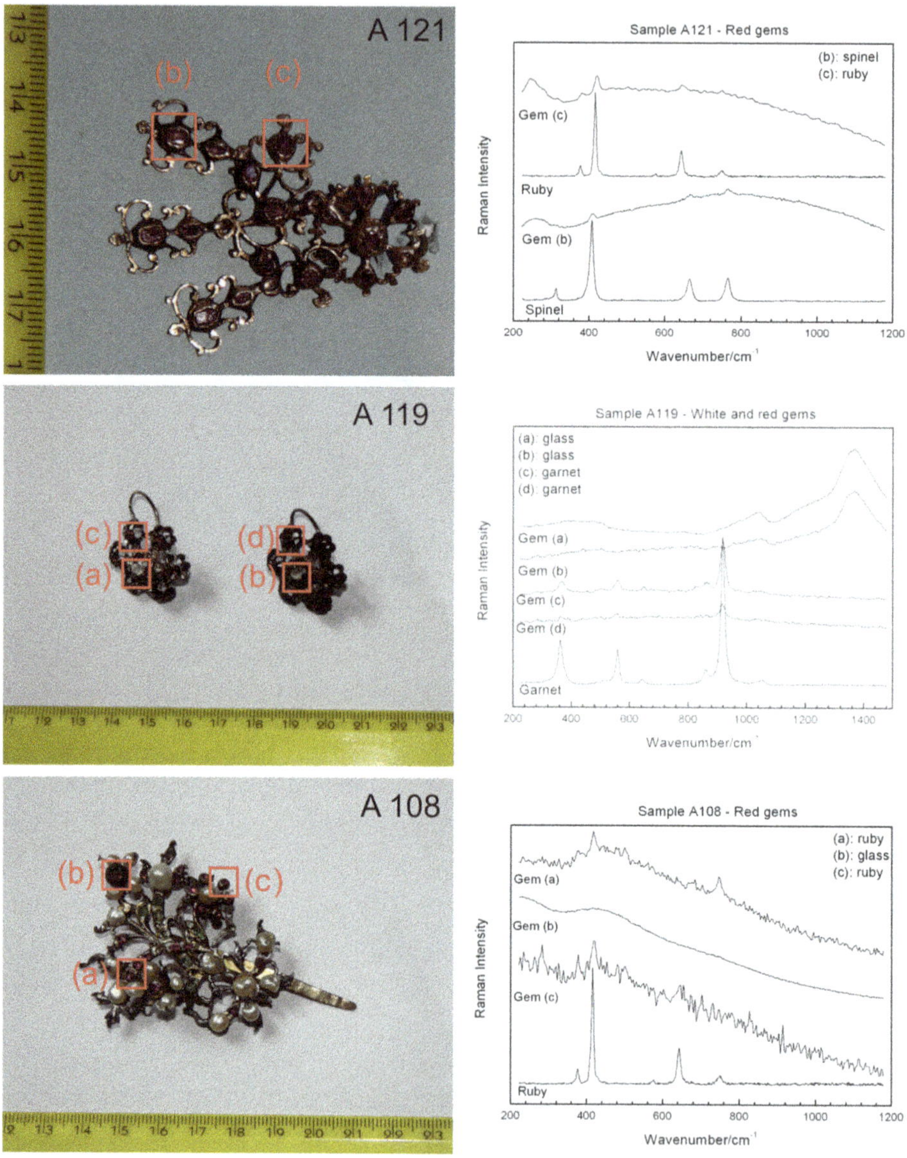

Figure 2.20. Pictures and Raman spectra acquired on jewels of the 17th–18th century collection of the Messina Regional Museum (Sicily, Italy), revealing rubies and rubies simulants: red spinel, garnet, and glass. The spectra were acquired using a 785 nm mobile Raman spectrometer. Adapted from [27] by permission of John Wiley & Sons, copyright 2015.

epoxy resins, and Canada balsams, have been identified by their 'fingerprint' in the spectral range 1200–1800 cm^{-1}, typical for organic compounds, where C–O and C–C stretching vibrational modes occur. Distinguishing the type of filler is of crucial importance in gemology, since some treatments are accepted (e.g. oils), whereas others are not allowed (e.g. epoxy resins), due to the difficulty of removing them caused by their polymerisation. Besides organic materials, fissures have also been filled using high-refractive index glass for brightening gems. Lead glass has been detected by Raman spectroscopy as a crack filler in rubies [21, 81]. An example of lead glass was previously shown in figure 2.12. In rubies, not-allowed treatments have been identified with micro-Raman detection of traces of flux agents on the surface of gemstones [85]. Other enhancement treatments changing the colour of gemstones by dyeing have been detected by Raman spectroscopy in freshwater pearls, as previously described in section 2.3.7 about the nature of colour in pearls and corals [3, 79].

Fundamental information about authenticity may also be obtained with mobile Raman investigations. Remarkably, handheld Raman measurements are achievable on gemological objects which cannot be moved from their location—e.g. from museums or sacred places—due to their high value [16, 23, 26, 27]. A mobile Raman campaign performed on several jewels belonging to the collection of the Messina Interdisciplinary Regional Museum in Sicily (Italy) successfully identified different types of forgeries, suggesting a different classification from the Raman attribution of the analysed gems [27]. In the 17th–18th century collection of the museum, besides the confirmation of the presence of rubies in some of the analysed jewels, various ruby simulants have been identified, consisting of red spinel, garnet, and glass (figure 2.20). Despite the fluorescence which is generally observed by using a 785 nm excitation in gems containing Cr^{3+} ions, the identification of rubies has been appropriately obtained by the mobile spectrometer. Interestingly, the garnet variety

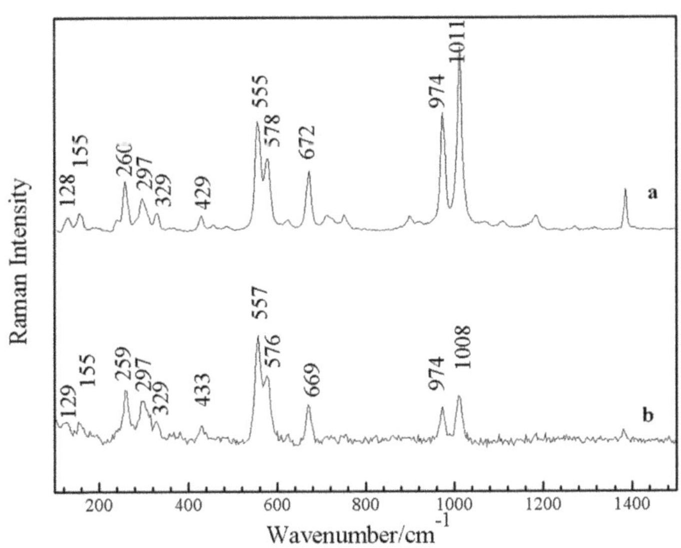

Figure 2.21. Raman spectra on sapphire simulant consisting of iolite, obtained with (a) micro-Raman at 473.1 nm and (b) mobile Raman equipment at 785 nm. Reproduced from [23] by permission of John Wiley & Sons, copyright 2014.

has been properly distinguished as almandine, though the spectral resolution of mobile devices is lower than that of laboratory equipment [31].

Another example of a simulant for sapphire is iolite, a transparent variety of cordierite ($Mg_2Al_4Si_5O_{18}$) used as cheaper gemstone, which has been identified by mobile Raman measurements (figure 2.21) [23]. Comparing the spectra obtained with mobile and micro-Raman spectrometers, it is noteworthy that most of the Raman features have been detected, including close-lying peaks which have been resolved.

Mobile Raman investigations are also effective in the identification of doublets or triplets. As in the case of the non-destructive campaign on the jewels belonging to the Messina Regional Museum, a doublet consisting of diamond and quartz was detected in a cross-pendant (figure 2.22(i)). Another recognition of a doublet was

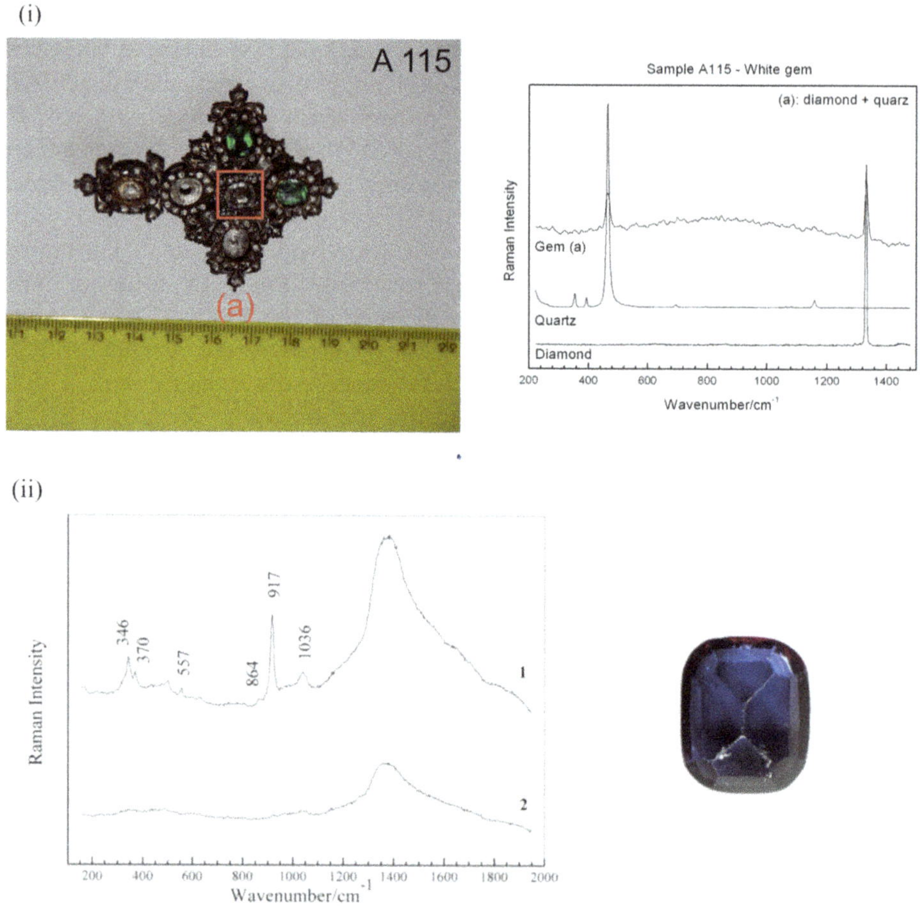

Figure 2.22. (i) Picture and Raman spectrum acquired on a jewel of the 17th–18th century collection of the Messina Regional Museum (Sicily, Italy), revealing a doublet with diamond and quartz. (ii) Picture and Raman spectrum of a garnet topped doublet with (1) garnet (on the top) and (2) glass (on the bottom). The spectra were acquired using 785 nm mobile Raman spectrometers. Reproduced from [23] and [27] by permission from John Wiley and Sons, copyright (2015), and by permissions of Regione Siciliana, Assessorato dei Beni Culturali e della Identità Siciliana, Dipartimento dei Beni Culturali e della Identità Siciliana – Museo Interdisciplinare di Messina, Italy.

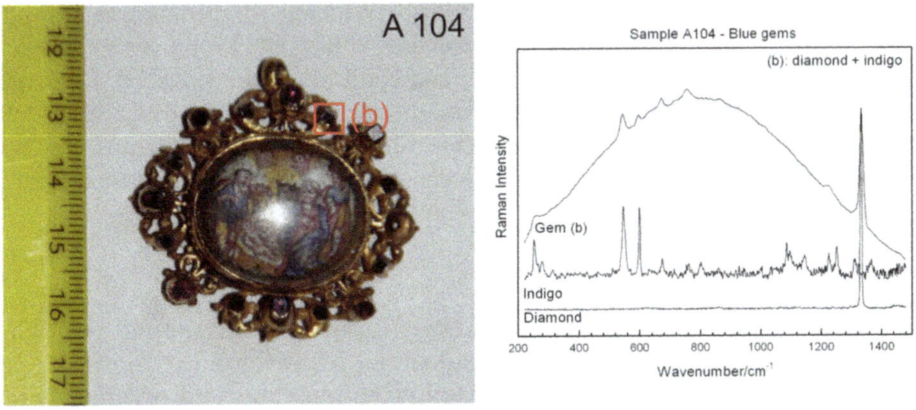

Figure 2.23. Picture and Raman spectra on a blue stone in a pendant, belonging to the 17th–18th century collection of the Messina Regional Museum (Sicily, Italy), revealing diamond (on the top) and indigo blue dye (on the bottom). The spectra were acquired using a 785 nm mobile Raman spectrometer. Reproduced from [27] by permissions of Regione Siciliana, Assessorato dei Beni Culturali e della Identità Siciliana, Dipartimento dei Beni Culturali e della Identità Siciliana – Museo Interdisciplinare di Messina, Italy.

obtained with mobile equipment on a garnet topped doublet: on the top of the sample the Raman spectrum of garnet was observed, whereas glass was found on its bottom (figure 2.22(ii)) [23, 27].

Distinguishing synthetic gemstones from their natural counterparts is also achievable with handheld Raman spectrometers. As previously described in section 2.3.3 about beryls, the band position and the width of the peak at ~1067–1068 cm^{-1} are indicative markers for synthetic gemstones, showing lower wavenumbers and a narrower width, respectively. In addition, in the OH-stretching spectral range, different 'water types' may be distinguished for low-alkali, high-alkali, or hydrothermal beryls. Furthermore, synthetic emeralds obtained by melt may be identified by the absence of OH-stretching peaks [16].

Interestingly, other expedients to obtain unusual chromatic effects on gems can be successfully distinguished with mobile Raman analyses. As proof of that, a blue stone of a pendant was demonstrated to consist of diamond and indigo blue dye, giving to the final object a grey-blue hue: in the upper part of the jewel, diamond was recognized by its characteristic intense Raman peak at 1333 cm^{-1}, while the bottom part was blue painted with indigo, as confirmed by its Raman signals (figure 2.23) [27].

2.4 Conclusions

Raman spectroscopy has proven to be one of the most useful techniques for the analysis of materials for gemological purposes. The ability to provide structural information in a non-destructive way makes Raman spectroscopy one of the first-choice techniques for analysing inorganic and organic gems. Moreover, the technique can be applied directly and on the field, with mobile Raman systems, without jeopardising the artefacts measured or the immovable gemological collections. Indeed, Raman spectroscopy has been applied in a broad variety of materials considered for gemological purposes, including garnets, jades, beryls/emeralds, sapphires, rubies, etc. Moreover, materials like glass and organic gems, for example pearls and corals, are also described. Forgeries of gemstones are included in Raman spectroscopy analysis and are described as glass, synthetics gems, plastics, etc among others. Finally, luminescence bands can reveal valuable details of the structure and the formation/genesis of the gemstones and can accompany the 'traditional' Raman procedure.

Acknowledgments

AR gratefully acknowledges Ghent University for its financial support through the Special Research Fund-BOF (BOF—postdoctoral fellowship).

The authors would like to thank the following museums and organisations for allowing us the study of their precious collections and the publication of their data: Soprintendenza Archeologia, Belle Arti e Paesaggio di Milano, Messina Interdisciplinary Regional Museum Maria Accascina (Messina, Italy), Messina Regional Museum (Sicily, Italy), Paolo Orsi Regional Museum (Siracusa, Italy), Museum LausPompeia (Lodi Vecchio, Lodi, Italy).

References

[1] Dhamelincourt P 1987 Laser Raman and fluorescence microprobing techniques *Anal. Chim. Acta* **195** 33–43

[2] Coupry C and Brissaud D 1996 Applications in art, jewelry and forensic science ed G Turrell and J Corset *Raman Microscopy.* (New York: Academic) pp 421–53

[3] Bergamonti L, Bersani D, Csermely D and Lottici P P 2011 The nature of the pigments in corals and pearls: a contribution from Raman spectroscopy *Spectrosc. Lett.* **44** 453–8

[4] Clark R J H 1995 Pigment identification microscopy *J. Mol. Struct.* **347** 417–28

[5] Clark R J H 1995 Raman microscopy: application to the identification of pigments on medieval manuscripts *Chem. Soc. Rev.* **24** 187–96

[6] Best S P, Clark R J H and Withnall R 1992 Non-destructive pigment analysis of artefacts by Raman microscopy *Endeavour* **16** 66–73

[7] Vandenabeele P, Verpoort F and Moens L 2001 Non-destructive analysis of paintings using Fourier transform Raman spectroscopy with fibre optics *J. Raman Spectrosc.* **32** 263–9

[8] Vandenabeele P, Weis T L, Grant E R and Moens L J 2004 A new instrument adapted to *in situ* Raman analysis of objects of art *Anal. Bioanal. Chem.* **379** 137–42

[9] Bersani D and Lottici P P 2010 Applications of Raman spectroscopy to gemology *Anal. Bioanal. Chem.* **397** 2631–46

[10] Vandenabeele P, Edwards H G M and Jehlička J 2014 The role of mobile instrumentation in novel applications of Raman spectroscopy: archaeometry, geosciences, and forensics *Chem. Soc. Rev.* **43** 2628–49

[11] Bersani D and Lottici P P 2016 Raman spectroscopy of minerals and mineral pigments in archaeometry *J. Raman Spectrosc.* **47** 499–530

[12] Bersani D, Conti C, Matousek P, Pozzi F and Vandenabeele P 2016 Methodological evolutions of Raman spectroscopy in art and archaeology *Anal. Methods* **8** 8395–409

[13] Rousaki A and Vandenabeele P 2021 In situ Raman spectroscopy for cultural heritage studies *J. Raman Spectrosc.* **52** 2178–89

[14] Jehlička J and Culka A 2021 Critical evaluation of portable Raman spectrometers: from rock outcrops and planetary analogs to cultural heritage – a review *Anal. Chim. Acta* **1209** 339027
[15] Rousaki A, Costa M, Saelens D, Lycke S, Sánchez A and Tuñón J *et al* 2020 A comparative mobile Raman study for the on field analysis of the *Mosaico de los Amores* of the Cástulo Archaeological Site (Linares, Spain) *J. Raman Spectrosc.* **51** 1913–23
[16] Jehlička J, Culka A, Bersani D and Vandenabeele P 2017 Comparison of seven portable Raman spectrometers: beryl as a case study *J. Raman Spectrosc.* **48** 1289–99
[17] Culka A and Jehlička J 2019 A database of Raman spectra of precious gemstones and minerals used as cut gems obtained using portable sequentially shifted excitation Raman spectrometer *J. Raman Spectrosc.* **50** 262–80
[18] Culka A and Jehlička J 2019 Identification of gemstones using portable sequentially shifted excitation Raman spectrometer and RRUFF online database: a proof of concept study *Eur. Phys. J. Plus.* **134** 130
[19] Coccato A, Bersani D, Caggiani M C, Mazzoleni P and Barone G 2021 Raman studies on zoisite and tanzanite for gemological applications *J. Raman Spectrosc.* **53** 550–62
[20] Bersani D, Azzi G, Lambruschi E, Barone G, Mazzoleni P and Raneri S *et al* 2014 Characterization of emeralds by micro-Raman spectroscopy *J. Raman Spectrosc.* **45** 1293–300
[21] Barone G, Bersani D, Lottici P P, Mazzoleni P, Raneri S and Longobardo U 2016 Red gemstone characterization by micro-Raman spectroscopy: the case of rubies and their imitations *J. Raman Spectrosc.* **47** 1534–9
[22] Casadio F, Douglas J G and Faber K T 2007 Noninvasive methods for the investigation of ancient Chinese jades: an integrated analytical approach *Anal Bioanal. Chem.* **387** 791–801
[23] Barone G, Bersani D, Crupi V, Longo F, Longobardo U and Lottici P P *et al* 2014 A portable versus micro-Raman equipment comparison for gemological purposes: the case of sapphires and their imitations *J. Raman Spectrosc.* **45** 1309–17
[24] Petrová Z, Jehlička J, Čapoun T, Hanus R, Trojek T and Goliáš V 2012 Gemstones and noble metals adorning the sceptre of the Faculty of Science of Charles University in Prague: integrated analysis by Raman and XRF handheld instruments *J. Raman Spectrosc.* **43** 1275–80
[25] Osterrothová K, Minaříková L, Culka A, Kuntoš J and Jehlička J 2014 *In situ* study of stones adorning a silver Torah shield using portable Raman spectrometers *J. Raman Spectrosc.* **45** 830–7
[26] Jehlička J, Culka A, Baštová M, Bašta P and Kuntoš J 2016 The Ring Monstrance from the Loreto treasury in Prague: handheld Raman spectrometer for identification of gemstones *Phil. Trans. R. Soc. A* **374** 2016004
[27] Barone G, Bersani D, Jehlička J, Lottici P P, Mazzoleni P and Raneri S *et al* 2015 Nondestructive investigation on the 17–18th centuries Sicilian jewelry collection at the Messina regional museum using mobile Raman equipment *J. Raman Spectrosc.* **46** 989–95
[28] Barone G, Mazzoleni P, Raneri S, Jehlička J, Vandenabeele P and Lottici P P *et al* 2016 Raman investigation of precious jewelry collections preserved in Paolo Orsi Regional Museum (Siracusa, Sicily) using portable equipment *Appl. Spectrosc.* **70** 1420–31
[29] Korsakov A V, Hutsebaut D, Theunissen K, Vandenabeele P and Stepanov A S 2007 Raman mapping of coesite inclusions in garnet from the Kokchetav Massif (Northern Kazakhstan), Spectrochim *Acta A: Mol. Biomol. Spectrosc.* **68** 1046–52
[30] Bernard S, Beyssac O and Benzerara K 2008 Raman mapping using advanced line-scanning systems: geological applications *Appl. Spectrosc.* **62** 1180–8

[31] Bersani D, Andò S, Vignola P, Moltifiori G, Marino I-G, Lottici P P and Diella V 2009 Micro-Raman spectroscopy as a routine tool for garnet analysis *Spectrochim. Acta A: Mol. Biomol. Spectrosc.* **73** 484–91

[32] Bersani D, Andò S, Vignola P, Marino I -G and Lottici P P 2009 Micro-Raman determination of the composition of ugrandite garnets *AIP Conf. Proc.* **1163** 35–43

[33] Hofmeister A M and Chopelas A 1991 Vibrational spectroscopy of end-member silicate garnets *Phys. Chem. Miner.* **17** 503–26

[34] Kolesov B A and Geiger C A 1998 Raman spectra of silicate garnets *Phys. Chem. Miner.* **25** 142–51

[35] Chopelas A 2005 Single crystal Raman spectrum of uvarovite, $Ca_3Cr_2Si_3O_{12}$ *Phys. Chem. Miner.* **32** 525–30

[36] Kolesov B A and Geiger C A 2000 Low-temperature single-crystal Raman spectrum of pyrope *Phys. Chem. Miner.* **27** 645–9

[37] Smith D C 2005 The RAMANITA© method for non-destructive and *in situ* semi-quantitative chemical analysis of mineral solid-solutions by multidimensional calibration of Raman wavenumber shifts *Spectrochim. Acta A: Mol. Biomol. Spectrosc.* **61** 2299–314

[38] Pinet M and Smith D C 1993 La microspectrométrie Raman des grenats $X_3Y_2Z_3O_{12}$: I. La série calcique naturelle ouvarovite-grossulaire-andradite *Schweiz. Mineral. Petrogr. Mitt.* **73** 21–40

[39] Pinet M and Smith D C 1994 La microspectrométrie Raman des grenats $X_3Y_2Z_3O_{12}$: II. La série alumineuse naturelle pyrope-almandin-spessartite *Schweiz. Mineral. Petrogr. Mitt.* **74** 161–79

[40] Walker J 1991 Jade: a special gemstone *Jade.* ed R Keverne (New York: Springer) 19–41

[41] Tribaudino M, Mantovani L, Bersani D and Lottici P P 2012 Raman spectroscopy of (Ca, Mg)$MgSi_2O_6$ clinopyroxenes *Am. Mineral.* **97** 1339–47

[42] Prencipe M, Mantovani L, Tribaudino M, Bersani D and Lottici P P 2012 The Raman spectrum of diopside: a comparison between *ab initio* calculated and experimentally measured frequencies *Eur. J. Mineral.* **24** 457–64

[43] Huang E, Chen C H, Huang T, Lin E H and Xu J-A 2000 Raman spectroscopic characteristics of Mg–Fe–Ca pyroxenes *Am. Mineral.* **85** 473–9

[44] Gendron F 2002 Discovery of jadeite-jade in Guatemala confirmed by non-destructive Raman microscopy *J. Archaeol. Sci.* **29** 837–51

[45] Bersani D, Andò S, Scrocco L, Gentile P, Salvioli-Mariani E, Fornasini L and Lottici P P 2019 Composition of amphiboles in the tremolite–ferro–actinolite series by Raman spectroscopy *Minerals* **9** 491

[46] Wang A, Dhamelincourt P and Turrell G 1988 Raman microspectroscopic study of the cation distribution in amphiboles *Appl. Spectrosc.* **42** 1441–50

[47] Waeselmann N, Schlüter J, Malcherek T, Della Ventura G, Oberti R and Mihailova B 2020 Nondestructive determination of the amphibole crystal-chemical formulae by Raman spectroscopy: one step closer *J. Raman Spectrosc.* **51** 1530–48

[48] Leissner L, Schlüter J, Horn I and Mihailova B 2015 Exploring the potential of Raman spectroscopy for crystallochemical analyses of complex hydrous silicates: I. Amphiboles *Am. Mineral.* **100** 2682–94

[49] Petriglieri J R, Bersani D, Laporte-Magoni C, Tribaudino M, Cavallo A, Salvioli-Mariani E and Turci F 2021 Portable Raman spectrometer for *in situ* analysis of asbestos and fibrous minerals *Appl. Sci.* **11** 287

[50] Petriglieri J R, Salvioli-Mariani E, Mantovani L, Tribaudino M, Lottici P P, Laporte-Magoni C and Bersani D 2015 Micro-Raman mapping of the polymorphs of serpentine *J. Raman Spectrosc.* **46** 953–8

[51] Kim C, Bell M I and McKeown D A 1995 Vibrational analysis of beryl ($Be_3Al_2Si_6O_{18}$) and its constituent ring (Si_6O_{18}) *Physica B* **205** 193–208

[52] Moroz I, Roth M, Boudelle M and Panczer G 2000 Raman microspectroscopy and fluorescence of emeralds from various deposits *J. Raman Spectrosc.* **31** 485–90

[53] Huong L T-T 2008 Microscopic, chemical and spectroscopic investigations in emeralds of various origins *PhD Thesis* University of Mainz

[54] Huong L T-T, Häger T and Hofmeister W 2010 Confocal micro-Raman spectroscopy: a powerful tool to identify natural and synthetic emeralds *Gems Gemol.* **46** 36–41

[55] Lambruschi E, Gatta G D, Adamo I, Bersani D, Salvioli Mariani E and Lottici P P 2014 Vibrational characterization of the new gemstone pezzottaite $Cs(Be_2Li)Al_2Si_6O_{18}$ and Cs-beryl *J. Raman Spectrosc.* **45** 993–9

[56] Wang W, Scarratt K, Emmett J L, Breeding C M and Douthit T R 2006 The effects of heat treatment on zircon inclusions in Madagascar sapphires *Gems Gemol.* **42** 134–50

[57] Long D 2002 *The Raman Effect: A Unified Treatment of the Theory of Raman Scattering by Molecules* (New York: Wiley)

[58] Karampelas S and Wörle M 2021 Spectroscopic study of the coloured gems in a 19th century pendant from Einsiedeln Abbey *J. Raman Spectrosc.* **53** 563–9

[59] Eaton-Magaña S, Breeding C M, Palke A C, Homkrajae A, Sun Z and McElhenny G 2021 Raman and photoluminescence mapping of gem materials *Minerals* **11** 177

[60] Colomban P, Tournie A and Bellot-Gurlet L 2006 Raman identification of glassy silicates used in ceramics, glass and jewellery: a tentative differentiation guide *J. Raman Spectrosc.* **37** 841–52

[61] Colomban P 2003 Polymerisation degree and Raman identification of ancient glasses used for jewellery, ceramics enamels and mosaics *J. Non-Crystalline Solids* **323** 180–7

[62] Cagno S, Van der Snickt G, Legrand S, Caen J, Patin M and Meulebroeck W *et al* 2021 Comparison of four mobile, non-invasive diagnostic techniques for differentiating glass types in historical leaded windows: MA-XRF, UV–Vis–NIR, Raman spectroscopy and IRT *X-Ray Spectrom.* **50** 293–309

[63] Colomban P, Etcheverry M P, Asquier M, Bounichou M and Tournié A 2006 Raman identification of ancient stained glasses and their degree of deterioration *J. Raman Spectrosc.* **37** 614–26

[64] Dirlam D M, Misiorowski E B and Thomas S A 1985 Pearl fashion through the ages *Gems Gemol.* **21** 63–78

[65] Smith C P, McClure S F, Eaton-Magaña S and Kondo D M 2007 Pink-to-red coral: a guide to determining origin of color *Gems Gemol* **43** 4–15

[66] Karampelas S, Kiefert L, Bersani D and Vandenabeele P 2020 *Gems and Gemmology: An Introduction for Archaeologists, Art-Historians and Conservators* (Berlin: Springer Nature)

[67] Hänni H 2012 Natural pearls and cultured pearls: a basic concept and its variations *Aust. Gemmol.* **24** 258–66

[68] Raneri S, Barone G, Mazzoleni P and Bersani D 2020 Non-destructive spectroscopic methods for gem analysis: a short review *2020 IMEKO TC-4 Int. Conf. Metrol. Archaeol. Cult. Herit* 501–6

[69] Wehrmeister U, Jacob D E, Soldati A L, Häger T and Hofmeister W 2007 Vaterite in freshwater cultured pearls from China and Japan *J. Gemmol.* **30** 399–412

[70] Bergamonti L, Bersani D, Mantovan S and Lottici P P 2013 Micro-Raman investigation of pigments and carbonate phases in corals and molluscan shells *Eur. J. Mineral.* **25** 845–53

[71] Karampelas S, Fritsch E, Mevellec J-Y, Gauthier J-P, Sklavounos S and Soldatos T 2007 Determination by Raman scattering of the nature of pigments in cultured freshwater pearls from the mollusk *Hyriopsis cumingi* *J. Raman Spectrosc.* **38** 217–30

[72] Barnard W and De Waal D 2006 Raman investigation of pigmentary molecules in the molluscan biogenic matrix *J. Raman Spectrosc.* **37** 342–52

[73] Maia L F, De Oliveira V E, Oliveira M E R, Reis F D, Fleury B G, Edwards H G M and De Oliveira L F C 2013 Colour diversification in octocorals based on conjugated polyenes: a Raman spectroscopic view *J. Raman Spectrosc.* **44** 560–6

[74] Merlin J C 1987 Resonance Raman analysis of astaxanthin–protein complexes *J. Raman Spectrosc.* **18** 519–23

[75] Kuzmany H 1980 Resonance Raman scattering from neutral and doped polyacetylene *Phys. Status Solidi* **97** 521–31

[76] Schaffer H E, Chance R R, Silbey R J, Knoll K and Schrock R R 1991 Conjugation length dependence of Raman scattering in a series of linear polyenes: Implications for polyacetylene *J. Chem. Phys.* **94** 4161–70

[77] Soldati A, Jacob D, Wehrmeister U, Häger T and Hofmeister W 2008 Micro-Raman spectroscopy of pigments contained in different calcium carbonate polymorphs from fresh-water cultured pearls *J. Raman Spectrosc.* **39** 525–36

[78] Kiefert L, Moreno D M, Arizmendi E, Hänni H and Elen S 2004 Cultured pearls from the Gulf of California, Mexico *Gems Gemol* **40** 26–38

[79] Kiefert L, Hänni H and Ostertag T 2001 Raman spectroscopic applications to gemmology *Pract. Spectrosc. Ser.* **28** 469–90

[80] Hänni H, Schubiger B, Kiefert L and Häberli S 1998 Raman investigations on two historical objects from Basel Cathedral: the reliquary cross and Dorothy monstrance *Gems Gemol* **34** 102–25

[81] Fan J-L, Guo S-G and Liu X-L 2009 The application of confocal micro-Raman spectrometer to nondestructive identification of filled gemstones *Spectrosc. Lett.* **42** 129–35

[82] Nassau K 2000 Gemstone materials *Kirk-Othmer Encyclopedia of Chemical Technology* (New York: Wiley)

[83] McClure S F and Smith C P 2000 Gemstone enhancement and detection in the 1990s *Gems Gemol* **36** 336–59 Available form

[84] Kiefert L, Haenni H A and Chalain J-P 2000 Identification of gemstone treatments with Raman spectroscopy *Optical Devices and Diagnostics in Materials Science* ed D L Andrews, T Asakura, S Jutamulia, W P Kirk, M G Lagally, R B Lal, J D Trolinger and D L Andrews (Bellingham, WA: SPIE Optical Engineering Press) pp 241–51

[85] Del Castillo H C, Deprez N, Dupuis T, Mathis F, Deneckere A, Vandenabeele P, Calderon T and Strivay D 2009 Towards the differentiation of non-treated and treated corundum minerals by ion-beam-induced luminescence and other complementary techniques *Anal Bioanal Chem* **394** 1043–58

[86] Frezzotti M L, Tecce F and Casagli A 2012 Raman spectroscopy for fluid inclusion analysis *J Geochemical Explor* **112** 1–20

[87] Dele M L, Dhamelincourt P, Poirot J P, Dereppe J M and Moreaux C 1997 Use of spectroscopic techniques for the study of natural and synthetic gems: application to rubies *J. Raman Spectrosc.* **28** 673–6

IOP Publishing

Artificial Intelligence and Spectroscopic Techniques for Gemology Applications

Ashutosh Kumar Shukla

Chapter 3

Application of Fourier-transformed infrared spectroscopy and machine learning algorithm for gem identification

Pimthong Thongnopkun, Kanet Wongravee, Prompong Pienpinijtham and Aumaparn Phlayrahan

3.1 Introduction

Fourier-transformed infrared (FTIR) spectroscopy has proven to be a valuable molecular spectroscopic technique to characterize gemstones due to the molecular specific fingerprints of the vibrational spectra. The spectral fingerprints provide information directly related to chemical structure, chemical compositions, impurities, and some additives in the stone. Moreover, FTIR spectrometer is easy to use and is a standard instrument in gemmological laboratories around the world. In this chapter, we present the theory of infrared (IR) spectroscopy. In addition, FTIR sampling techniques for gemstone analysis are discussed. Although various FTIR accessories designed for specific applications and sample form are available, the technique encounter limitations. Moreover, examples of utilizing an FTIR approach for significant commercial gemstones (i.e., diamond, ruby and sapphire, emerald, jade, quartz, and turquoise) are demonstrated. At the end of this chapter, we introduce an application of machine learning algorithm on gemstone classification, which will pave the way for high-throughput characterizations of gemstones in the near future.

3.2 Concept of IR spectroscopy

IR spectroscopy is a kind of molecular vibrational spectroscopy which uses electromagnetic waves with wavelengths in the IR region (0.7–1000 μm wavelength) as a light source. Naturally, molecules are driven to vibrate (change bond length or

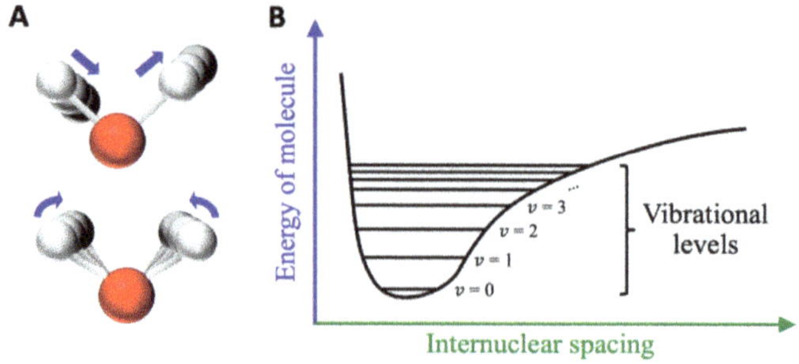

Figure 3.1. (a) Examples of molecular vibrations of an H_2O molecule: stretching (a change in bond length) and bending (a change in bond angle). (b) The relation between internuclear spacing and energy of a diatomic molecule.

bond angle, as depicted in figure 3.1(a)) by a surrounding temperature. However, the molecules can also absorb IR light, and then the vibrational states of the molecules are changed (molecules vibrate faster or wider). We can consider the internuclear spacing and energy of a diatomic molecule, as presented in figure 3.1(b). When two nuclei of atoms move close to each other (the internuclear spacing is reduced) to form a chemical bonding, the energy of the system (two nuclei) decreases. As mentioned above, the vibrational state of the molecule can be changed from the state $\nu = 0$ to the state $\nu = 1$ by absorbing an external energy, e.g., electromagnetic radiation or light. Coincidentally, the energy gap between those two states matches to the energy of mid-IR (2.5–25 μm wavelength). Moreover, the energy gaps between the state $\nu = 0$ to other states correspond to the energy of near-IR radiation (NIR; 0.7–2.5 μm wavelength).

Different types of chemical bonds (different pairs of atoms) have their own unique energy gaps as described, which cause different amounts of energy (different frequencies/wavelengths of IR light) to be absorbed by the bonds. Furthermore, different compounds are composed of different chemical bonds. Therefore, an IR absorption pattern (IR spectrum) is characteristic of each compound, which can be used to distinguish it from others. By this reason, IR spectrum is well known as a 'fingerprint' of compounds.

The interaction between IR light and matters normally can be observed by three different ways: absorption, reflection, and transmission, which lead to many different measurement modes of IR spectroscopy such as transmission mode, reflection mode, diffuse reflectance mode, attenuated total reflection (ATR) mode, etc. Nevertheless, the main concept of IR measurement is the illumination of IR light onto a sample with the detection of unabsorbed IR light from a sample (reflected light and transmitted light). The two main modes (setups) of IR spectroscopy, i.e., transmission and reflection modes, are illustrated in figure 3.2. It is noted that other measurement modes are subtypes of these two. For the transmission mode (figure 3.2(a)), the sample in this measurement mode must be thin and allow IR light to transmit. IR light is illuminated on a sample (I_{incident}). The sample absorbs some of the IR light while unabsorbed IR light transmits through the sample to a

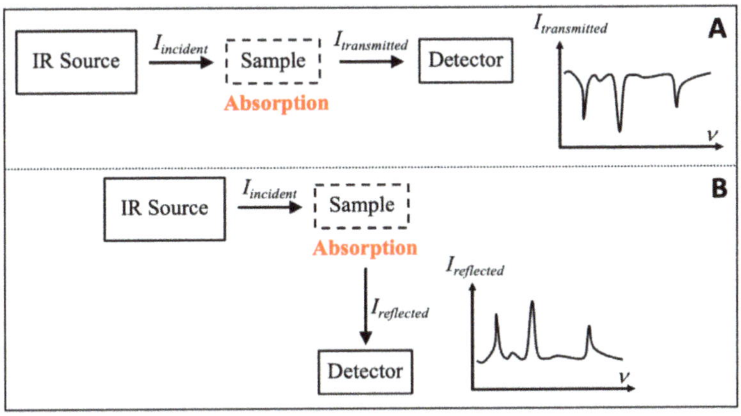

Figure 3.2. (a) Transmission and (b) reflection measurement modes of IR spectroscopy.

detector ($I_{\text{transmitted}}$). The plot between the intensity of the transmitted light and IR wavelength/frequency is a so-called 'transmission spectrum.' On the other hand, the sample for the reflection mode might be opaque for IR light, but it must have a surface with good reflection of IR light. In figure 3.2(b), the unabsorbed IR light reflects from the sample surface and goes to the detector ($I_{\text{reflected}}$). The plot between the intensity of the reflected light and the wavelength/frequency is known as 'reflection spectrum.' Both transmission spectrum and reflection spectrum can also be converted to 'absorption spectrum,' which presents the meaning of which wavelength/frequency is absorbed by the sample.

Other measurement modes have then been developed later after those two measurement modes to collect the spectrum from samples with various physical appearances which are not suitable to be operated by transmission and reflection modes. For example, a stiff opaque sample with a grain shape or a rough surface cannot be prepared for transmission measurement, and it cannot also reflect IR light well to a detector.

To increase a scanning speed of a spectral collection, an IR spectrometer has been integrated with a 'Michelson interferometer'; the spectrometer does not collect the signal wavelength by wavelength (or frequency by frequency). The Michelson interferometer improves the spectrometer to collect all wavelengths of signals together and record the IR intensity as a function of time. Then, the signals are converted using a 'Fourier transform' to generate an IR spectrum (IR intensity as a function of wavelength/frequency). Therefore, the name of the technique has been specified as FTIR.

One convenient FTIR sampling technique is an ATR mode. This mode helps researchers deal with various types of samples that are difficult to record a spectrum by a common transmission or reflection mode. ATR mode is based on a phenomenon of total internal reflection (TIR). Ordinarily, the light travelling from a higher-refractive-index medium to a lower-refractive-index medium refracts with the refracted angle which is larger than an incident angle. The incident angle

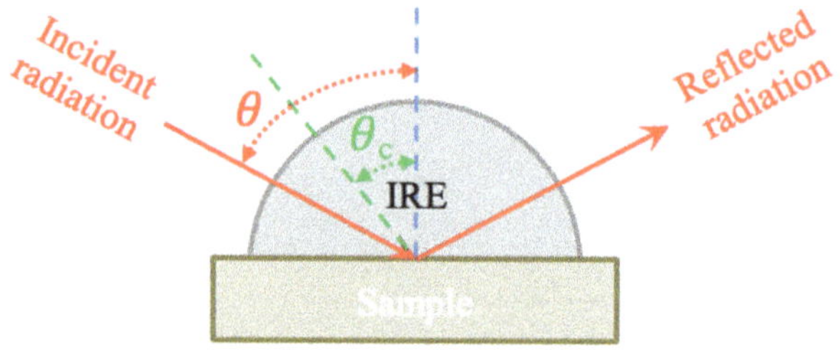

Figure 3.3. Light path in ATR measurement.

that causes the refracted radiation going parallel to an interface between two media is defined as the 'critical angle (θ_c).' When the incident angle is larger than a critical angle, there is no light going across the interface. Hence, the light undergoes TIR by being reflected to the higher-refractive-index medium with the reflected angle which equals an incident angle. The higher-refractive-index medium is usually called the 'internal reflection element (IRE)'. If the lower-refractive-index medium can absorb any wavelength/frequency of light, the intensity of light will decrease because an evanescence wave is generated during TIR and penetrates the lower-refractive-index medium. When the IRE is placed on a sample with a low refractive index and IR light is illuminated through the IRE to the sample, the sample can absorb IR light, and unabsorbed IR light can be reflected to a detector (see figure 3.3). This is the mechanism of ATR mode to record an IR spectrum, and the IRE is sometimes called 'ATR prism.'

There are three criteria to select materials for fabricating an ATR prism used in ATR accessories. The first one is that the material must be inert enough to not react with samples. The second is that the material must allow IR light to travel through it. The last one is that the material must possess a refractive index higher than samples. There are currently many available materials for making ATR prism, e.g., AMTIR, diamond, Ge, KRS-5, silicon, ZnS, and ZnSe.

ATR mode can be applied to a wide range of samples, e.g., cotton fabrics [1], polysaccharides [2], membranes [3], photocatalysts [4], thin films [5], polymers [6], medicines [7], vitamins in fruits [8], sugar in honey [9], silk sericin protein [10], etc. This mode can be performed with both liquid and solid samples. There is no problem with a variation of sample shapes. It is also a non-destructive measurement method, with no complicated sample preparation required before measurement. The only important key for ATR mode is that the optical contact between ATR prism and sample must be obtained. Accordingly, there are several ATR accessories developed to serve various kinds of samples with different purposes of measurements.

Nowadays, ATR mode has been developed and coupled with an IR microscope to collect a spectrum of microscale samples; the technique is known as 'ATR-FTIR microspectroscopy.' To accomplish this setup, the shape of ATR prism has been redesigned from hemispherical/triangular prism/trapezoid prism to dome-shaped. By redesigning the shape of ATR prism, a spectrum of a tiny samples such as a piece of microplastics [11], a fiber [12], or even a single human hair [13] can be measured at

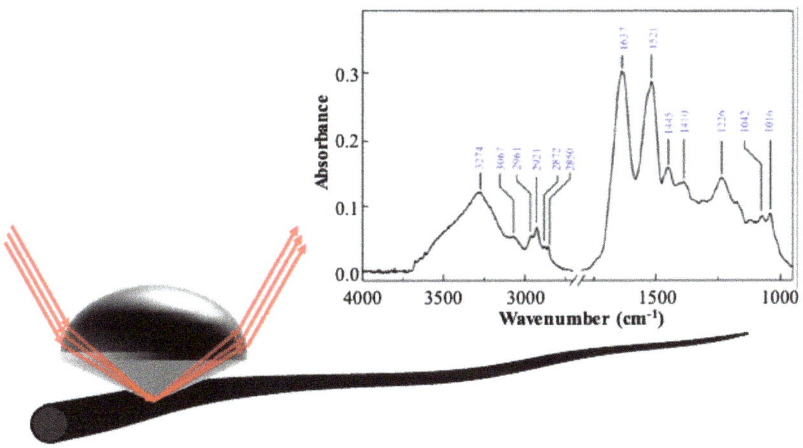

Figure 3.4. A dome-shaped Ge μIRE was employed to measure IR spectrum of a single human hair [13] (reproduced from reference [13] with permission).

specific positions. For example, Pienpinijtham *et al* presented a use of ATR-FTIR microspectroscopy to investigate a single human hair. In this work, ATR prism was fabricated to be dome-shaped using Ge as an IRE base material, as presented in figure 3.4. A homemade dome-shaped Ge μIRE accessory was installed on an IR microscope. The positions for IR measurement were selected specifically by looking through the IR microscope before making an optical contact between the ATR prism and the sample. IR spectra of single human hairs and cosmetic residues on them can be characterized selectively at microscale levels. The differences between un-split and split hairs can also be observed in the IR spectra of this work. It is a good example that ATR-FTIR microspectroscopy can be exploited with a broad range of samples or even very small samples that are quite difficult to handle using a conventional IR spectroscopy.

One of the most popular and convenient modes for IR measurement of a gemstone is diffuse reflectance IR Fourier transform spectroscopy (DRIFT). In general, this technique is a common sampling technique for the study of powders, solids, and species absorbed on solids. Nowadays, the diffuse reflectance technique is very popular for gem analysis because it is easily used and non-destructive. The theory of diffuse reflectance based on incident radiation on a sample being scattering in all direction or the process which the angle of reflection is different from the angle of incidence. The spectrum will be expressed in terms of a Kubelka–Munk unit. Although the diffuse reflectance spectrum is very similar to the transmission spectrum, the relative intensities of the absorption bands are different. DRIFTS is a powerful non-destructive technique for gemstone analysis. However, the accessory and its sample holder has a restricted working area which cannot perform for a large sample. Other drawbacks of the diffuse reflectance technique arise from the specular reflection component in which the component does not penetrate the sample. The spectral results will reveal Reststrahlen bands, which is a particular problem for interpretation.

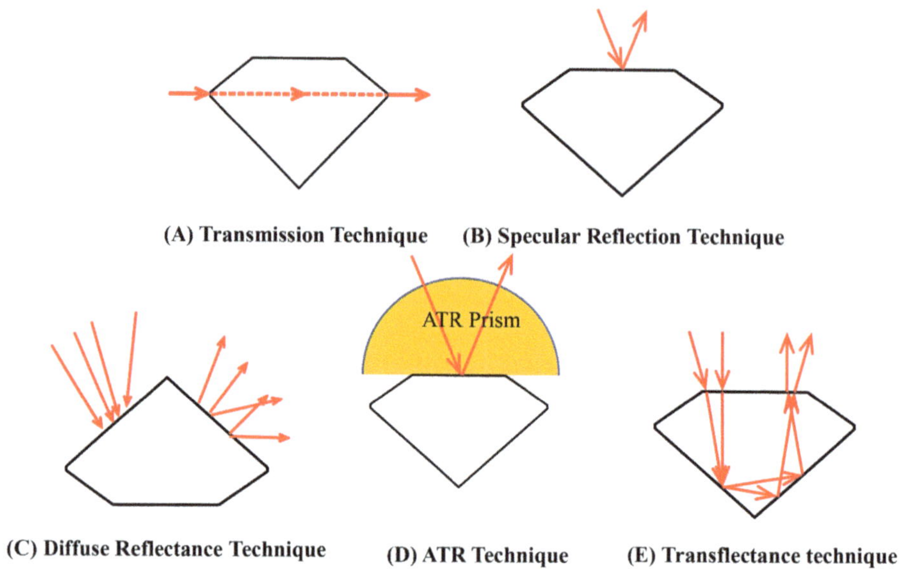

Figure 3.5. Schematic diagram of FTIR sampling techniques applied for faceted gemstone analysis: (a) transmission, (b) specular reflection, (c) diffuse reflection, (d) ATR, and (e) transflectance technique.

Transflectance technique is a sampling technique that is introduced for the characterization of faceted gems and gems on jewellry [14, 15]. The technique is based on the principle of reflection and refraction within gemstones. The term 'transflectance' phenomena comprises the essential part of optics: reflection, refraction, and critical angle. When light attacks the surface of a well-polished gemstone, some of that incident light is *reflected* back from the surface, and some remaining incident light is *refracted* and *transmitted* through the surface of a gemstone and then approached to the detector, as shown in figure 3.5. In principle with gem cutting, a faceted gemstone is cut in such a proportion that the TIR within the gemstone is enhanced. This principle can help to produce transflectance phenomena in gem samples. In order to collect transflectance spectra, this technique has to be used with an IR microscope or specific designed spectrometer. The technique is non-destructive and can solve the problems associated with the characterization of faceted gemstone and jewellery [14].

3.2.1 FT-IR sampling techniques for gem analysis

Nowadays, FT-IR spectroscopy is widely employed for gem classification based on chemical composition and impurities in the crystal lattice. Since various FT-IR sampling techniques and accessories are designed for specific application and sample form, different sampling techniques encounter different difficulties. Figure 3.5 presents the schematic diagram of FTIR sampling techniques that was applied for faceted gemstone measurement.

Transmission technique: For classical gemstone characterization using the transmission technique, a thin slab of gems or a KBr pellet method is required. Because the technique is destructive, it is not suitable for high commercial value gemstones such as faceted diamonds, ruby, sapphire, emerald, etc. A beam condenser is an accessory for measuring diamond spectrum via the transmission method, but the measurement of a faceted diamond using a beam condenser is complicated by sample arrangement (figure 3.5(a)) and the complex reflection at the cut and polished surface.

Specular reflection: Specular reflection is defined as the reflection at which the ange of incident radiation is exactly equal to the angle of reflection, as shown in figure 3.5(b). It is particularly useful for the characterization of faceted gems, but it is necessary for the surface of gems to have a flat polished facet and be very smooth. If the gem surface is not smooth or it has a curved or spherical surface, it is very difficult to obtain a high quality of reflection spectrum. The obtained spectrum may not be identified. Nevertheless, the specular reflection sampling technique is a very rapid and non-destructive determination of the sample. The technique is an available non-destructive technique for the characterization of opaque gemstones such as turquoise. The KK-transformed spectrum agrees very well with the transmission spectrum [14, 15]. The specular reflection spectrum, on the other hand, does not show any prominent absorption feature of diamond because of its high reflective index.

DRIFT: In order to solve the problems associated with the characterization of faceted gemstones, the diffuse reflectance technique is required for the measurement. The diffuse reflection diagram to measure faceted gems is shown in figure 3.5(c). Currently, the diffuse reflectance technique is the most convenient sampling technique that has been successfully employed for IR spectral acquisition of the faceted gemstones because of its non-destructive nature, short measurement time, and simplicity. However, the sampling technique has both benefits and limitations. Since most gems are anisotropic and the spectrum depends on crystal orientation, the measurements of faceted gemstones tend to be non-reproducible. The diffuse reflectance spectra of a gemstone are greatly influenced by the gemstone arrangement on the sample holder [14–17]. Moreover, the technique is not applicable for mounted gems on a jewellery body due to a limitation of sample holder size. Besides, the technique is not suitable for cabochon shape gemstones [14].

ATR technique: ATR FT-IR spectroscopy is one of the powerful characterization techniques using IR radiation. The technique is very rapid and requires very small sample amounts and virtually no sample preparation. In addition, high-performance ATR accessories provide a simple and effective alternative to the traditional and time-consuming technique. To collect ATR spectrum of faceted gemstones, the sample was placed under an ATR prism as illustrated in figure 3.5(c). For the spectral acquisition of faceted gems, the single-reflection ATR technique is more applicable for spectral acquisition than the multiple reflection because of the small and hard surfaces of the sample. There are various types of ATR prisms such as ZnSe, Ge, Si, KRS-5, and ZnS; all of these crystals have different refractive indexes and spectral ranges. The ATR prism should have a higher index of refraction than the sample. In the case of diamond, the prism suitable for diamond analysis are germanium (Ge) or silicon because they have higher refractive index than that of diamond (2.417). For diamond analysis, it can provide diamond information, and the obtained ATR spectrum agreed with the well-accepted diffuse reflectance spectrum [18]. Nevertheless, ATR may not be a suitable technique for diamond analysis due to some limitations of spectral acquisition. The ATR spectral acquisition must be very careful because the crystal surface can be damaged by the extremely hard surface of faceted diamond. An air gap between the ATR crystal and a diamond sample can deteriorate the spectral quality. If the air gap is thick enough, an ATR FT-IR spectrum cannot be observed. ATR technique has been applied for the analysis of other gemstones such as emerald, turquoise, and amber [14, 19]. For amber analysis, the diamond ATR prism was employed for investigation. FTIR spectra of ambers are useful to distinguish amber from immature resins such as copal and to determine the origin of archaeological ambers [19]. Although ATR-FTIR has great advantages for gemstone identification, the technique is not applicable for analysing mounted diamonds on jewellery body.

Transflectance technique: The transflectance technique via an IR microscope is an alternative method which is applicable to acquire IR spectra of loose and faceted gemstones. The diagram of transflectance technique applied on gemstone analysis was shown as in the figure 3.5(e). The technique is non-destructive, provides unambiguous results, requires no sample preparation, and has a short analysis time. Moreover, the distinct advantage of the transflectance technique over the diffuse reflectance or transmission techniques is its ability to acquire IR spectra of mounted gemstones on the jewellery without taking the gemstone out of the jewellery setting [14, 15, 20]. The observed transflectance spectrum shows the same spectral features as those of the well-accepted diffuse reflectance spectrum [14, 15]. In addition, unlike the diffuse reflectance spectrum, the transflectance spectrum is not altered by the sample arrangements [14, 15]. Nevertheless, the technique encounters some limitations that depend strongly on the shape and size of sample. The maximum thickness that can be performed with the transflectance technique is approximately 8 mm. The technique cannot collect transflected radiation from poor-cut proportion gemstones [14]. Unfortunately, the technique cannot be used for opaque gemstones such as turquoise, ivory, pearl, etc.

Therefore, the appropriated sampling technique is very important for investigating materials, including faceted gemstones in order to obtain high spectral quality. In the next topic, examples of the FTIR approach for some commercial gemstones (i.e., diamond, ruby and sapphire, emerald, jade, quartz, and turquoise) are discussed.

3.3 Diamond

FT-IR spectroscopy is a powerful spectroscopic technique that is widely employed for diamond classification based on impurities (i.e., nitrogen, hydrogen, and boron) in the diamond crystal lattice. Most of the FTIR applications on diamond analysis are the classification of the type of diamond, differentiation between natural and synthetic diamonds, and determination of the treatment processes applied onto the diamonds. In addition, the technique is also used for differentiating diamond simulants from natural diamonds [14, 15].

3.3.1 Classification of diamond types

Diamond is one of the most precious gemstones that has long been the prime interest in gems and jewellery trade due to its popularity, mystic symbolism, and commercial value. Because of its significance, developments on diamond identification have introduced many scientific concepts into the gemmological work. Diamond 'type' is one of the most important concepts frequently mentioned in the gemmological area. The classification system of diamond type is widely used in diamond research because it is an accessible way to categorize diamonds based on their defects or impurities in the crystal lattice. Understanding diamond types is critical to evaluating the relationships between diamond growth, synthetic process, treatment process, and color.

It is well known that the chemical composition of diamond is a carbon atom in a regular repeating pattern. In fact, some atoms of elements such as nitrogen (N) and boron (B) can replace carbon atoms in the lattice of diamond, and other impurities can be incorporated. The diamond type classification was divided into categories based on the presence and absence of nitrogen impurities that are arranged in the diamond lattice [21]. Diamonds can be broadly divided into two main types, namely type I and type II.

Figure 3.6 presents the classification of diamond type based on nitrogen and boron impurities in the lattice. Type I diamond is defined as containing sufficient N

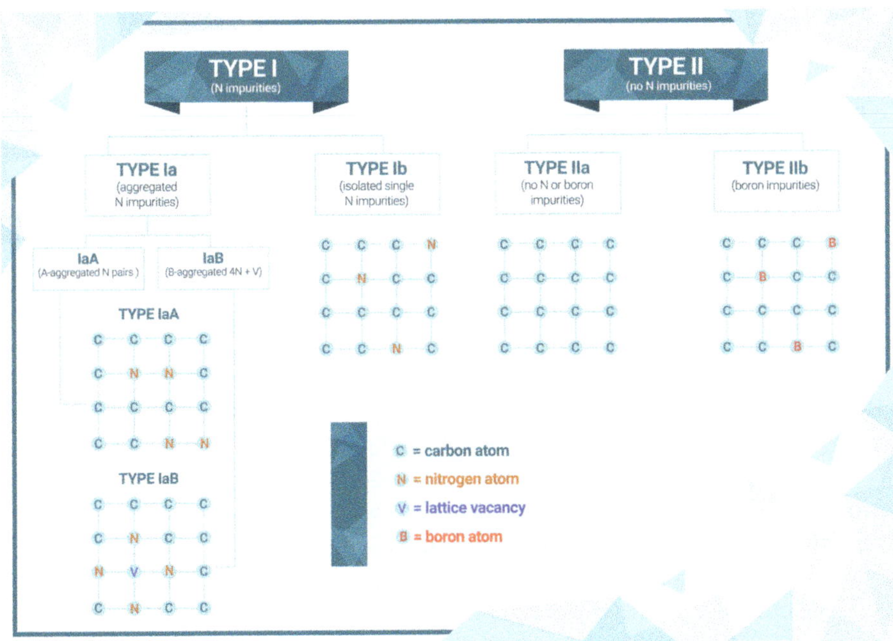

Figure 3.6. Diamond type classification is based on nitrogen and boron impurities in the lattice.

to be investigated by FTIR spectroscopy. With respect to the arrangement of nitrogen atoms, type I diamond is subdivided into type Ia (aggregated nitrogen atoms) and type Ib (single substituted nitrogen atoms). In addition, type Ia diamond is also separated into type IaA and type IaB depending on nitrogen aggregates in its lattice. Type IaA has nitrogen pairs known as A-centers, while type IaB has rings of four nitrogen atoms symmetrically surrounding a vacancy known as B-centers. The degree of nitrogen aggregation is dependent on the concentration, the time, and the residence temperature of diamonds in Earth's mantle.

Type Ia diamond is the most common configuration for natural diamond, but pure type Ib diamond is very rare in nature. In type Ib diamond, carbon atoms in lattices are placed by a single nitrogen atom, and the nitrogen atom is isolated from another one known as C-center. When diamonds crystallize, C-center is the original form of incorporated nitrogen impurities in the diamond lattice. The C-center can migrate and combines to form A-center after the diamonds reside in the mantle for sufficient time. As the aggregation conducts, A-center further migrates and combines to form B-centers. However, the complete transition from A-center to B-center is rarely achieved because the process requires a high activation energy. Therefore, pure type IaB diamonds are uncommon. The diamonds instead usually contain significant concentrations of both A-centers and B-centers, known as mixed type IaAB. In type II diamond, it does not contain enough nitrogen atoms that can detected by FTIR spectroscopy (around 5–10 ppm). Type II diamond is subdivided into type IIa (absence of boron impurity) and type IIb (single substituted boron atoms) [21–24].

For natural diamond, a vast majority of natural diamonds are type Ia, approximately 98% in proportion (both colorless and fancy color), followed by type IIa diamonds, with a population of less than 2%. There is a lack of type Ib and type IIb diamonds in nature, about less than 0.1% of all-natural diamonds [21–24]. An FTIR spectrometer is the most important instrument to identify the type of diamonds [21]. It is useful to quantitatively analyse the concentration and configuration of nitrogen atoms in diamonds that were classified in figure 3.6. Occasionally, diamond type can provide a few clues for differentiating between the natural and synthetic stone [21, 22].

To discuss the type of diamond and their impurities using FTIR spectroscopy, the detail of a diamond spectrum must first be described. Figure 3.7 reveals examples of the FTIR spectra of diamond acquired by diffuse reflectance technique. The diamond spectrum shows the important absorption band in the mid-IR region (4000–400 cm^{-1}). For the FTIR spectrum of diamond, the spectrum range was divided into three zones based on the chemical bonds between carbon atoms or carbon and any impurities within the diamond lattice when exposed to IR radiation. The three zones are called the one-, two-, and three-phonon regions [21]. The one-phonon region is the range of about 1330 to 400 cm^{-1}, which is related to the characteristic absorption of nitrogen (N) impurity. The two-phonon region (2665–2330 cm^{-1}) and three-phonon region (4000–2665 cm^{-1}) shows important

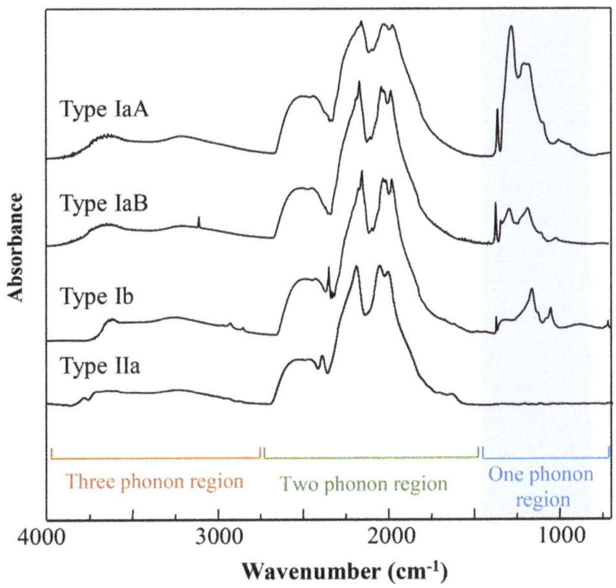

Figure 3.7. The FTIR spectrum of diamond consists of one-, two-, and three-phonon regions. (A) type IaA, (B) type IaB, (C) type Ib, and (D) type IIa.

features intrinsic to diamond. The two-phonon region is caused by vibration of carbon–carbon bonds of the diamond lattice. In addition, boron impurity can be easily detected at ~2458 cm^{-1} in the two-phonon region and at ~2930 and ~2800 cm^{-1} in the three-phonon region. This area also revealed other impurities such as hydrogen that do not affect the type classification [21–24].

According to the IR spectra of diamond in figure 3.7, the area that is important to separate between subgroups of diamond is the one-phonon region. Type I diamond illustrates the distinctive spectral features related to different configurations of nitrogen impurities in the range of about 1330 to 400 cm^{-1} (the one-phonon region). As shown in that region, the diamond spectra in figure 3.7(a) were identified as a type IaA, while that in figure 3.7(b) was assigned as a type IaB. Considering figure 3.7(a), the unique absorption bands of the type IaA diamond possesses an A-center defect where the main absorption has a maximum at 1282 cm^{-1} and additional weak absorptions at 1362 cm^{-1}, 1208 cm^{-1}, and 1015 cm^{-1}. The unique absorption bands in the one-phonon region of the type IaB diamond associates with a B-center defect showing the absorption band at 1330 cm^{-1}, 1172 cm^{-1}, and 1013 cm^{-1}. An additional absorption band at 1360 cm^{-1} referred to the extended-planar-defect platelets. In addition, minor spectral features commonly observed in type IaB diamond are a strong absorption at 3107 cm^{-1} and a weak absorption at 2786 cm^{-1}. The bands correspond to the absorption of a hydrogen–nitrogen vibration. It should be noted that variable concentrations of A and B aggregates along with single substitutional N create a continuum of peak intensities in this region. In the case of type Ib (figure 3.7(c)), the appearance of N impurities from isolated or single N can be detected at 1344 and ~1130 cm^{-1} [15, 21]. The spectrum also revealed absorption at 3107 cm^{-1} and 2450 cm^{-1} which related to hydrogen impurities. If the intensity of absorption at 3107 cm^{-1} is very high and greater than that of absorption at 2450 cm^{-1}, the diamond is called 'Hydrogen rich' (H-rich). IR absorption of hydrogen in natural diamond was investigated by Fritsch *et al* [23].

For type IIa in figure 3.7(d), there are no detectable impurities in the one-phonon region that can be investigated by FTIR spectroscopy. Thus, the absorption of type IIa diamond reveals only absorption of the two-phonon region, intrinsic to diamond character. For type IIb, the diamond is a blue color because of boron impurity. The IR spectrum of natural type IIb diamond shows boron-related peaks at 2459 cm^{-1} and 2802 cm^{-1} [21]. The color of diamond correlates to the impurities in diamond and also diamond type. Characteristic absorption band and peak assignments of natural diamond spectrum corresponding to defects in diamond and diamond type are summarized in table 3.1.

Table 3.1. Characteristic bands and assignments corresponding to diamond type.

Type of diamond	Peak position indicator	Impurity	Most common color
Ia	Aggregated nitrogen at 1282 cm^{-1}, 1175 cm^{-1}	Nitrogen (Aggregated)	**Natural diamond**: colorless, yellow, orange, green, brown, pink **Treated diamond**: yellow, orange, red, blue
IaA	A-center (maximum at 1282 cm^{-1} and weak additional absorptions at 1362 cm^{-1}, 1208 cm^{-1}, and 1015 cm^{-1})		
IaB	B-center defect (maximum at 1175 cm^{-1} and additional absorptions at 1330 cm^{-1}, 1172 cm^{-1}, 3107 cm^{-1}, and 2786 cm^{-1-})		
Ib	Isolated nitrogen at 1344 cm^{-1}, 1130 cm^{-1}	Nitrogen (isolated)	**Natural diamond**: yellow, brown **Irradiation and annealed diamond**: pink to red
IIa	No detectable impurity	None	**Natural diamond**: colorless, pink, brown, green **HPHT-treated diamond**: enhance brown to colorless or pink
IIb	Boron at 2459 cm^{-1}, 2802 cm^{-1}, 2928 cm^{-1}	Boron	**Natural diamond**: blue, gray **HPHT-treated diamond**: enhance blue

3.3.2 Characterization of synthetic diamonds and treated diamonds

3.3.2.1 Synthetic diamonds

Since natural diamonds have been exposed during their extended period in the Earth and the conditions imposed in a laboratory during treatment or synthetic growth are quite different, the impurities in lattice defects of natural diamond are different from the diamond evolution during lab growth or treatment. Therefore, the impurities in diamonds or diamond type are very important for separating natural and synthetic diamonds, and adequate understanding of diamond type is critical for identification purposes [25–31].

Over the last several years, the production of high-pressure high-temperature (HPHT)-grown synthetic diamond has increased dramatically [27–30]. The common type of natural diamond is type Ia, while HPHT-grown synthetic diamonds are almost all type Ib [21, 32, 33]. As mentioned above, type Ib is very rare in natural. Those few natural diamonds that are type Ib usually contain abundant inclusions of natural minerals and exhibit colorful strain patterns. However, type Ib HPHT-grown synthetic diamonds contain only metallic flux inclusions. Nowadays, most synthetic type Ib diamonds in the market have an intense yellow to orange color.

Nowadays, the chemical vapor deposition (CVD) synthetic diamond has flooded gem markets around the world. For CVD synthetic diamonds, they are most commonly type IIa, a type that is rare in natural diamonds. Thus, colorless type IIa diamond must be considered as a possible synthetic diamond. Occasionally, both HPHT and CVD growth techniques can produce type IIb blue diamond [21, 32–37]. However, the presence of electrical conductivity and the absence of tatami strain patterns can be utilized for differentiating the natural from synthetic diamond. For fancy colored diamond, the color of CVD diamond is caused by a strong absorption of very intense silicon-related SiV^- center (negatively charged silicon and vacancy center). It reveals a high concentration of Bi and Pb that was found by laser ablation inductively coupled plasma mass spectrometry (LA-ICP-MS). However, the non-destructive and effective identification methods of CVD are absorption bands from photoluminescence spectroscopy [26]. When compared to HPHT, the CVD diamond is simpler and has more reproduction capability than the HPHT diamond. Thus, CVD has the potential of mass produced cheap synthetic diamond. Besides its technological efficiency, CVD diamonds have the potential to control and produce color diamonds, which is very important in jewellery application [32–39].

CVD and HPHT synthetic processes can produce fancy color diamond and colorless diamond [32–41]. In the case of CVD-grown diamond, it can produce various colors, normally brown to colorless type IIa diamond. Zaitsev *et al* have examined various shades of yellow CVD-growth diamond and irradiated- and annealed-treated CVD-growth diamond produced by Scio Diamonds by spectroscopic techniques. The result showed that the intense yellow color of Scio Diamonds is due to nitrogen content in the diamond. Their FTIR spectra show small peaks at 1344 cm^{-1} and 1332 cm^{-1}, which are the characteristics of type IIa diamond.

Occasionally, brown high-growth-rate CVD diamond can show the broad band at 2930 cm^{-1} attributed to hydrogenated amorphous carbon. In addition, a small peak may be seen in the region of 2800–3200 cm^{-1} related to hybridized carbon and H-bonds from the synthetic and annealing processes.

In the case of fancy color diamond, natural pink diamonds, especially of an intense color, are very rare and in high demand in the diamond market. Most natural pink diamonds come from the Argyle Mine, Australia. Nowadays, pink CVD-grown diamond is produced from several producers and brought onto the market. The natural and synthetic diamonds from several companies were investigated by Deljanin *et al* [38]. The results showed that most natural pink diamonds are type IaAB with a platelet peak at 1362 cm^{-1} and also display the hydrogen-related peak (H-center) at 3107 cm^{-1}, 4494 cm^{-1}, and 1405 cm^{-1}. Type IIa natural pink diamonds are quite rare in nature (less than 2% of all-natural pinks). When comparing the FTIR spectra of pink diamond from Arglye Mine and non-Argyle Mine, the characteristic or fingerprint of the Arglye Mine is that the absorption of B-aggregate is higher than that of A-aggregate [25, 39]. In the case of CVD synthetic, most pink CVD-grown diamonds are type IIa. It reveals the main feature of spectra of all pink CVD diamonds at 1292 cm^{-1}, which originate from defects common for these diamonds. Moreover, the diamonds also reveal small peaks at 1503, 1450 (H1a center), 1340 and 1332 (N^+ center), 1295 and 1128 (C centers), and 1130 (single nitrogen), but they have negligible absorption in the range 1000 to 1500 cm^{-1} [38].

For blue diamond, the natural diamond is type IIb and shows boron-related peaks at 2459 cm^{-1}, 2802 cm^{-1}, and 2928 cm^{-1} (also with small peaks at 1332 cm^{-1} and 1303 cm^{-1}), while the CVD diamonds are type IIa and present weak bands at 1340 cm^{-1} and 1332 cm^{-1}. All HPHT synthetic blue diamonds are type IIb, which has high concentrations of Fe and Co that can be detected by EDXRF [32]. For the IR spectrum, the HPHT must show very strong peaks of boron at 2455 cm^{-1} and 2802 cm^{-1} and be oversaturated above 2750 cm^{-1} [33]. In the case of boron-doped CVD diamond, main absorption peaks at 520 cm^{-1} and 1282 cm^{-1} are observed. The bands related to boron carbon vibration (B–C) have been identified as substitutional boron in diamond. It also shows the band at 1160 cm^{-1} that was assigned to B–O [40].

3.3.2.2 Treated diamonds

Beside the synthetics, various types of treatments (i.e., irradiation, heat, low pressure–high temperature, annealing, HPHT treatments) or annealing or a combination of these processes were applied to the low-quality diamonds in order to enhance their color and appearances (i.e., inclusion elimination, color enhancement, and clarity improvement) [21, 41–45]. Since most natural diamonds show a brown hue in color and some internal or external defect, considerable treated techniques have been continuously made to enhance the color of diamonds to satisfy the demand for color gem-quality diamond. In some cases, rapid grown CVD diamonds with a brown or greyish brown color can considerably reduce the undesirable brown color by HPHT treatment [41–44]. Due to the same chemical structure and composition, the treated and synthetic diamonds are more difficult to detect or to exclude out of the natural untreated counterparts by the basic gemmological techniques (i.e., refractive index, hardness, color, specific gravity, and light dispersion measurements). Thus, the characterization of them becomes more critical to gem customers. The large price difference between the synthetic or treated diamonds and natural untreated diamonds is very important for gemmologists to correctly identify the diamonds. The gemmological and spectroscopic properties have been combined for conclusive diamond identification.

For heat-treated fancy colored diamond, FTIR spectroscopy is used as a preliminary screening technique for the separation of treated pink diamonds from the natural one based on their type. However, photoluminescence and visible spectroscopy are necessary for reliable identification. For HPHT-treated pink diamond, the diamond can be type Ia and type IIa or type Ib depending on the process of each producer. HPHT synthetic, irradiated Chatham, and Russia must be type IIa with a small peak of H1a (1450 cm^{-1}) and N^+ (1332 cm^{-1}). Moreover, it presents C-center peaks at 1344 cm^{-1} and 1129 cm^{-1} and small peaks around 1560 cm^{-1} (1560 cm^{-1}, 1545 cm^{-1}, 1537 cm^{-1}, 1504 cm^{-1}, 1451 cm^{-1}) with a side band at 1470 cm^{-1}. In the case of type Ib, the spectrum indicated the C-center (1344 cm^{-1} and 1130 cm^{-1}) and a series peak at 1560 cm^{-1}, 1502 cm^{-1}, 1450 cm^{-1}, and 1332 cm^{-1} [34, 35, 38].

In case of irradiation-treated diamond, most of the diamonds are assigned to type IaAB. The characteristic absorption of the H1a defect at 1450 cm^{-1} is observed in all irradiation diamonds, which can indicate that the diamond has been irradiated or irradiated with subsequent annealing with low temperature (300 °C). In case of blue diamond, the band at 2800 cm^{-1} disappeared, which is evidence for the irradiation existence. Moreover, the peak located at 1356–1380 cm^{-1} related to C–C platelets, by product of nitrogen aggregation, was presented in the irradiated diamond. Besides, the peak at 2785 cm^{-1} and double peak at 2926 cm^{-1} and 2845 cm^{-1} are also observed [41–45].

3.3.3 Identification of diamond imitations

Diamond has been imitated by artificial materials. A diamond simulant or diamond imitation is a material that has a different chemical composition with similar appearance to a diamond. Advanced technology has been developed for increasing better simulants with properties near those of diamond. Since diamond simulants do not share the same chemical compositions with natural diamond, the trained gemmologists can distinguish diamond from all diamond simulants primarily by visual inspection and basic gemmological equipment. The most common diamond simulants are cubic zirconia (ZrO_2), colorless synthetic corundum (sapphire; Al_2O_3), synthetic spinel, strontium titanate ($SrTiO_3$), yttrium aluminium garnet, gadolinium gallium garnet, and synthetic moissanite [29–31].

Although some diamond simulants can be identified by standard gemmological equipment, others cannot. Moreover, it is difficult to identify the type of simulant, especially in cases of diamond settings in a jewellery body. FTIR of diamond simulants have been investigated by diffuse reflectance technique and transflectance technique by Pimthong *et al* [15]. The spectral features of all simulants are obviously different to natural and synthetic diamond. The unique spectral feature of diamond in a two-phonon region cannot be occurred in all simulants. All simulants show saturation absorption in the low wavenumber and the weak absorption band at the high wavenumber region. Therefore, FTIR spectra can be easily differentiated between diamond and diamond simulants and also the identification of the type of diamond simulants based on their chemical composition.

3.4 Rubies and sapphires

3.4.1 General information on rubies and sapphires

3.4.1.1 What are ruby and sapphire?

In the gemstone world, ruby and blue sapphire are arguably classified as two of the most sought-after colored gemstones, considerably for their outstanding characteristics, i.e., beauty, durability, and rarity, as well as the long history attached to them. The stones are commonly known as gem-corundum. There are two varieties of gem-corundum. The stone possesses a red color called ruby, while the other colored stones are named sapphires, which can be in various ranges of a spectrum, i.e., violet, purple, green, yellow, orange, sapphires.

Ruby and sapphires are crystalline alpha-aluminum oxide (α-Al_2O_3) minerals. In its purest state, corundum is a colorless crystal. The Al_2O_3 forms an octahedral structure. Each aluminum ion is bonded with six oxygen ions arranging in a close-packed hexagonal. In 1925, the position of aluminum and oxygen ions in a corundum structure was closely re-determined by Linus Pauling and Sterling B. Hendricks [46]. The result indicated that the oxygen ions slightly distort and bond in a rhombohedral structure rather than in the octahedral one. The distance between aluminum ions with the three-upper position of oxygen ions is longer than those of the other three oxygen ions, approximately 0.145 Å with 60° of angle.

Although the gem-corundum possesses almost the same chemical composition and structure, a few percentages of trace elements can make them completely different in appearance. The impurity of Cr^{3+} replaces some positions of Al^{3+} in the Al_2O_3 crystal, causing the red color in ruby, while the inter-valent charge transfer between Fe^{2+} and Ti^{4+} influenced the blue color in sapphire. In addition, the Fe ions seem to play a significant role in creating green (Fe^{2+}) and yellow (Fe^{3+}) colors like those of the other sapphires.

The market price of ruby and sapphires can vary at many levels depending on their quality. In the past, color and clarity were the main factors to consider when determining the values. However, there are many more factors to concentrate on in the present. The synthetic materials, the enhancement, and the geographical origins are affecting the per-carat prices of ruby and sapphires.

3.4.1.2 Synthetic ruby and sapphire

Synthetic gemstones are human-made materials. The crystal is grown in the laboratory using the same chemical ingredients as its natural counterpart, thus showing similarity in virtual, chemical, optical, and physical properties. Synthetic ruby was the first material that could be produced in the faceting quality in the late 1800s. These materials are used in other industries such as laser technology, microelectronics, abrasive, and communications. For gem application, the synthetic stones can be custom made (controlling sizes and shapes), and thus, the values of these materials are relatively low compared to the natural stones. Therefore, the synthetic gem material should be disclosed as such at the time of sale to avoid any deceptions affecting the prices [47].

Synthetic corundum can be produced by a great number of processes. The common types of synthetic methods for ruby and sapphires are melt and solution. The melting processes include flame fusion or Verneuil process, crystal pulling, or Czochralski process. Flame fusion is the method started by melting the chemical powder via a high-temperature flame. The chemical composition used in the corundum synthetic process is Al_2O_3 with the colored elements, i.e., Cr_2O_3, TiO_2, FeO, Fe_2O_3. Then, the melted compound falls on a rotating platform and grows a crystal. On the other hand, flux-growth crystals can be synthesized by adding flux, a solid material dissolving the started chemical composition. The dissolved solution subsequently cooled down and formed a crystal. Synthetic ruby and sapphire crystals are widely available at various ranges of price levels. Nowadays, flame fusion, flux growth, and hydrothermal are still promising methods to produce good quality synthetic crystals.

3.4.1.3 Heat treatment of ruby and sapphire

Heat treatment is a potential method to enhance the quality of ruby and sapphire. This technique can be dated back a thousand years and was performed by the Sri Lankans. One of the first stones reported that underwent heat treatment was ruby. In the beginning, the heat treatment was simple. The technique involves a 'man blow' or blowpipe technique, which is when a heater blows a flame through the pipe. The temperature can reach around 1064 °C (melting point of gold). It is normally performed to remove the bluish tints from ruby and pink sapphire.

Nowadays, the treatments are more sophisticated than in the early years. The list of heat treatments for ruby and sapphire is shown in table 3.2 [48].

The other enhancement techniques used to improve the quality of ruby and sapphire, though uncommon, include irradiation, dyeing, and oiling.

3.4.1.4 Geographic origins of ruby and sapphire

In recent years, the shift of the ruby and blue sapphire supply chain from Asia to Africa and especially to Madagascar and Mozambique has inevitable changed the selling and buying methods in the gem market. The gemmological laboratories around the world face demand for certification that can tell the customer where the gem comes from. Normally, the value of the gems could be considered based on their beauty, rarity, and durability. However, provenance plays an increasingly important role in determining the value of the gems. There are various sources of ruby and sapphire around the world relating to different geological settings and metamorphic and magmatic processes [48].

Table 3.2. Heat treatments for ruby and sapphire.

Enhancement type	Purpose of heat treatment	Applied to	Price drop from natural of equal quality
Heat or heat and gas			
Heat to 800 °C–1200 °C in oxidizing atmosphere	Remove blue tints	Ruby, pink sapphire	Modest
Heat to 1400 °C–1800 °C in reducing atmosphere	Improve clarity and increase blue color	Blue sapphire	Significant
Heat to 1400 °C–1800 °C in oxidizing atmosphere	Develop yellow color	Yellow, orange sapphire	Significant
Heat with chemicals (lattice diffusion)			
Heat to 1600 °C–1900 °C with titanium in reducing atmosphere	Increase blue color (Ti diffusion)	Blue sapphire	Very significant
Heat to 1600 °C–1900 °C with beryllium	Increase yellow color (Be diffusion)	Ruby, pink, green, yellow sapphire	Very significant
Heat to 1600 °C–1900 °C with beryllium	Reduce blue color	Dark blue sapphire	Very significant
Heat with flux (flux healing)			
Heat to 1200 °C–1600 °C with flux	Heal the surface-reaching fracture	Mostly ruby	Significant
Heat with color glass impregnation			
Use acid to remove impurity in fracture, then impregnate the stone with high-refractive index colored glass	Improve clarity and color	Ruby and blue sapphire	Hugh

The major commercial old and new sources of ruby in the world are Mozambique, Madagascar, Tanzania, Myanmar, Vietnam, Kenya, Malawi, Pakistan, Tajikistan, Sri Lanka, and Thailand. Some sources are still continually active in terms of production, i.e., Mozambique, Madagascar, and Myanmar. The others are moving past their peak and produce extremely limited gems in today's trade, i.e., Afghanistan and Thailand [48].

The blue sapphire has been mined mainly from two geological occurrences, basaltic relates and metamorphic process. The most famous and historical blue sapphire came from Ceylon (now known as Sri Lanka). Nowadays, blue sapphire among other fancy colored sapphire is still unearthed in Sri Lanka mines. Kashmir sapphire is one of the most sought after colored stones in the world. The low-quality blue sapphire from this provenance can possess a value higher than those of the fine quality stone from other origins. Thailand, Cambodia, Australia, the United States of America, Madagascar, and East Africa (Tanzania, Kenya, Malawi) are also significant suppliers of blue sapphire [48].

3.4.2 Application of FITR for corundum analysis

3.4.2.1 Synthetic corundum

The 3309 cm^{-1}-series peaks in IR absorption at 3310 cm^{-1}, 3232 cm^{-1}, and 3185 cm^{-1} are one of the significant absorption peaks when studying corundum (Al_2O_3). The peaks seem to appear in all kinds of samples, whether natural, synthetic, or enhancement stones. In the history of applying IR spectroscopic techniques to identify corundum, these groups of peaks have always been mentioned [49–51].

At first, IR spectroscopy was practiced on synthetic corundum because it is easier to find the cause of the peak occurring when one knows what one detects. A certain amount of ingredients has been put together to create a synthetic material. In the case of corundum, Al_2O_3 was the main component in the synthesis methods along with a limited number of coloring atoms, i.e., Cr, Fe, V, Mg, Ni, and Co [52–57].

The FT-IR spectra of hydrothermal synthetic ruby showed the normal IR absorption of corundum in the range of 800 cm^{-1} to 400 cm^{-1}, which is also found in the other synthetic methods of corundum. In addition, these groups of peaks around the fingerprint region also appear in natural corundum [58]. The strongest absorption present in this region near 1700 cm^{-1} is assigned to Al–O stretching in the lattice absorption [49].

Hydrothermal growth crystal usually consists of water in different forms. The same characteristic happens in natural growth corundum [51] and can be found in Verneuil (flame-fusion) synthetic [50, 59]. Water as the OH^- group entered the corundum structure in both natural crystal and synthetic material to act as a charge compensation for Al^{3+}. In natural corundum, the OH^- group appears in the form of vacancies of Al in the octahedron structure [60]. In the synthetic material, the compensation of OH^- when added to the divalent cations, i.e., Fe^{2+}, Ni^{2+}, Co^{2+}, Mg^{2+}, is as coloring atoms into the Al_2O_3 structure [50, 59]. Greenish-blue and green synthetic sapphire showed the FT-IR spectra at the 4000–1600 cm^{-1} region. The wide band with the strongest absorption was at 2980 cm^{-1}, the narrow peak at 2750 cm^{-1}, and the fine narrow peak around 2500–2000 cm^{-1}. The greenish-blue synthetic sapphire showed the absorption band at 3300 cm^{-1} as a shoulder. The green color presents the wide bands at 3400 cm^{-1} and 3310 cm^{-1} in correlation intensity. These features are also observed in the FT-IR spectra of natural corundum containing divalent cations [61]. The Ni-doped hydrothermal synthetic sapphire and Verneuil synthetic process showed similar features in the OH^- regard. The samples also showed a notable peak at 2980 cm^{-1} defined as OH-group for charge compensation of $Al^{3+} + O^{2-} \rightarrow Ni^{2+} + (OH)^-$. The narrow peak between 3600 cm^{-1} and 3000 cm^{-1} is also related to the absorption of a group of OH^- compensation during the formation of cation vacancies in the Al_2O_3 structure, which can be found in both natural and synthetic corundum [50].

The absorption peak related to CO and CO_2 is one of the groups of peaks normally found in both synthetic and natural corundum. The C–O-related peaks usually appear around 2460 cm^{-1}, 2400 cm^{-1}, 2264 cm^{-1}, and 2016 cm^{-1}. The C–O band is caused by the corundum structure itself, not from impurity, i.e., CO or CO_2 fluid inclusion [57].

The Verneuil synthetic sapphire showed narrow peaks at 3310 cm^{-1}, 3232 cm^{-1}, and 3185 cm^{-1}. It can be separated into two types of absorption. The first is categorized by the strongest absorption peak at 3230 cm^{-1} along with the weak band at 3310 cm^{-1} and 3185 cm^{-1}. This type I peak could normally be found in colorless, rose red, and red ruby synthetic sapphire. The second type of peak can be detected in violet, deep red, light blue, and blue synthetic sapphire. It is a type II when the strongest absorption appears at 3309 cm^{-1}, and weak bands appear at 3230 cm^{-1} and 3185 cm^{-1}. Meanwhile, an additional peak at 3290 cm^{-1} could be found in the FT-IR spectra of the alexandrite-like synthetic sapphire which is defined as V^{3+} incorporation in Al^{3+}. The colorless and rose-ruby colored showed weak absorption bands at 3280 cm^{-1} and 3160 cm^{-1}. The 3280 cm^{-1} showed maximum absorption when the spectra were measured at the perpendicular direction of the *c*-axis [50].

The series of peaks at 3309 cm^{-1}, 3230 cm^{-1}, and 3185 cm^{-1} occur in both colorless and colored varieties of synthetic corundum. It is indicated that the splitting of the peak is dependent on the presence of vacancies in the cations lattice. The intensity of the peaks is controlled by the variable replacement mode of doped atoms in the channel of octahedron Al^{3+} cation coordination [50].

Moon and Phillips [62] intensively studied the defect structure of Ti^{4+} in α-Al_2O_3. The 73 ppm of tetravalent titanium is doped into a pale pink sapphire rod. The hydrogen protons were introduced to the structure by heating the crystal with H_2N_2 gas at 1873 K for 2 h. The set of sharp bands was detected in the IR absorption spectra of the Ti: α-Al_2O_3 synthetic pink sapphire at 3309 cm^{-1}, 33232 cm^{-1}, and 3187 cm^{-1}, along with the weak stretching frequency at 3368 cm^{-1} and 3295 cm^{-1}. The group of OH bands is assigned to a defect cluster of doped Ti atoms for charges compensating with cation vacancies and OH^- in the two adjacent Al atoms in an octahedron structure. However, the different stretching frequencies of the peaks indicate the different localities of the vacancies in the structure. The intensities of the OH band at 3309, 3232, 3187, 3368, and 3295 cm^{-1} have correlated with one another at a ratio of I_{3232}/I_{3187} around 3, I_{3368}/I_{3309} around 1/18, and I_{3295}/I_{3309} around 1/12. The reduced intensity was also shown in a proportional ratio after heating. The OH band demonstrated that H protons are associated with vacancies in Al and form an OH band with oxygen atoms, which is a co-ordinate in the charge vacancy. The H, Fe, Ti: α-Al_2O_3 of blue and blue–green synthetic sapphire also showed a set of stretching frequencies around 3000 cm^{-1} with the centered bands at 3309 cm^{-1}, 3232 cm^{-1}, 3187 cm^{-1}, 3367 cm^{-1}, and 3295 cm^{-1}. However, the half-width (FWHM) of these bands in H, Fe, Ti: α-Al_2O_3 were larger than those of H, Ti: α-Al_2O_3. The OH bands appear after hydrogen protons are incorporated into the crystal structure via either natural or hydrogen saturation processes. The intensity of the bands is related to the concentration of Ti atoms added to the structure. These sets of OH bands seem to be found only in the samples with doped Ti and Fe + Ti while disappearing in the samples containing only Fe [63].

Even so, FT-IR spectroscopy is potentially useful for identifying synthetic corundum. However, it should be carefully interpreted when using this technique to distinguish natural and synthetic corundum because some characteristics can occur in both varieties.

3.4.2.2 Heat treatment detection

The dominated absorption bands of Al–O stretching frequencies at 760, 642, 602, and 405 cm^{-1} are a main indicator of verifying the corundum. These peaks typically show strong absorption in the range below 1500 cm^{-1} [50]. However, a similar feature can also be observed in synthetic and heat-treated corundum. Natural unheated sapphire usually shows an absorption peak at 3309 cm^{-1} in varying intensities. Typically, the single peak at 3309 cm^{-1} is a shred of evidence for unheated blue sapphire. The small peak at 3232 cm^{-1} could develop after heat treatment and is strongly suggested as a characteristic of heat treatment in ruby and sapphire [48]. The unheated rubies, especially those from Mong Hsu, Myanmar, usually contain hydroxyl inclusion, i.e., diaspore (α-AlOOH), which is related to the occurrence of the absorption bands in FT-TR spectra around 1900 cm^{-1} to 3400 cm^{-1}. However, after heat treatment, the structurally OH absorption peaks of approximately 3100–3600 cm^{-1} play a significant role in heat treatment determination [51]. The absorption peaks around 2100–1980 cm^{-1} of OH stretching related to the diaspore mineral inclusion could be one of the clues to indicate the heat treatment at low temperatures (below 1200 °C). The peaks would slightly shift to a higher wavenumber after heated samples [64]. The extremely weak peak of OH absorption at 3310 cm^{-1}, 3232 cm^{-1}, and 3185 cm^{-1} can be present in the FT-IR spectra of ruby from Sri Lanka and blue sapphire from Montana. The absorption peaks around 1900 cm^{-1} and 2040 cm^{-1}, along with a broad absorption band between 2120 cm^{-1} and 2145 cm^{-1} and the band at 2885 cm^{-1} and 3025 cm^{-1}, can be found in ruby samples. These groups of bands were defined as OH absorption from different forms of water, i.e., H_2O, OH^-, and hydroxyl minerals. The intensity of the bands was prominently measured along the vibrational direction parallel to the *c*-axis of the Al_2O_3 structure. The features disappeared in the FT-IR spectra of corundum without water molecules or hydroxyl groups in the structure. The IR absorption of boehmite (γ-AlOOH) mineral inclusions was also detected in the ruby samples at 3080 cm^{-1} and 3290 cm^{-1} [50]. However, the alteration of the peaks was weak, and we could not observe the alteration during heat treatment. The increasing intensity of absorption peaks around 3600 cm^{-1} to 3100 cm^{-1} is normally higher when the samples are heated. The set of sharp peaks related to structural OH groups at 3310 cm^{-1}, side peaks at 3232, 3187, and 3368, and a shoulder peak at 3292 cm^{-1} were introduced after heat treatment. The appearance of the OH-related peaks depends on the different types of samples. Some samples show the absorption peaks only at 3309 cm^{-1}, while others may exhibit the 3309 cm^{-1} peaks together with their side peaks. Smith [51] also reported that the heated ruby samples usually show a structural OH group which is the result of the alteration and dehydration of hydroxyl mineral inclusion, particularly diaspore. The concentration of structural OH bands depends on the amount of this hydrous mineral inclusion in the 'host' ruby structure.

A strong absorption peak at 3310 cm^{-1} usually appears together with its side peaks at 3232 cm^{-1} and 3185 cm^{-1}, and only when the 3309 cm^{-1} is at its strongest, the additional weak peaks at 3395 cm^{-1} and 3367 cm^{-1} can appear. The color of the samples plays an important role in determining the intensity of the OH-related peaks. The blue sapphire exhibits the highest intensity, while those of the light blue sapphire are normally lower. The OH band suggested that it related to the redox reduction involving iron, which is opposed to the one concluded by Moon and Phillips (1994). Some studies also reported that the soaking time during the heat treatment process affects the intensity of the OH bands. The longer the soaking time, the weaker the intensity of the OH peaks [66].

3.4.2.3 Cobalt and glass-filled corundum

The FT-IR spectra can also be used to detect the filling corundum. The IR spectra of cobalt-doped glass-filled sapphire show wax peaks at 2620 cm^{-1}, 2579 cm^{-1}, 2290 cm^{-1}, 2256 cm^{-1}, 3490 cm^{-1}, and 3500 cm^{-1}, which is usually found in glass-filled ruby. The lead glass-filled corundum shows a strong absorption band at approximately 3480 cm^{-1} and 2618 cm^{-1}. In green sapphire, the IR spectrum usually contains the peaks at 3400 cm^{-1}, 2597 cm^{-1}, and 2256 cm^{-1}. These peaks are all related to the glass-filled materials [67–70].

3.4.2.4 Origin determination

In recent times, geographic origin determination has been one of the important factors when customers decide to buy rubies and sapphires. It has become a challenging issue that modern geological laboratories must encounter. For example, the value of a ruby from Myanmar can be worth ten times more than a ruby of the same quality from Madagascar. Meanwhile, low-quality blue sapphire from Kashmir possesses a price per carat higher than those of the better ones from Ceylon (Sri Lanka) based only on provenance. Thus, the gem industry, especially the gemmological laboratory, has to act quickly to procure a reliable method for the declaration of origin determination.

The 3309-cm^{-1} series peaks and the 3160-cm^{-1} series peaks are still the focused characteristics for origin determination. However, FT-IR spectroscopy for geographic origin determination is still in an exploratory stage. Natural sapphire from basaltic origin typically exhibits the OH band attributed to metal ions (Ti, V, and Fe) Ti–OH, V–OH, and Fe–OH. Meanwhile, the metamorphic sources do not show evidence of the OH group and very weak, if any, absorption. In blue sapphire samples, the OH band can be found in both basaltic and metamorphic origins. The basaltic origins normally indicate a 3309 cm^{-1} strong intensity, with the side peaks at 3232 cm^{-1} and 3185 cm^{-1}, while those of the metamorphic origin are only present in a relatively weak peak at 3309 cm^{-1} [65]. These features can be helpful in separating and grouping the samples into two main geological occurrences: basaltic related and metamorphic origins.

3.5 Emerald

Emerald is the green king of the four precious gemstones that has an irreplaceable position in the gems market. It is a green variety of the mineral beryl ($Be_3Al_2Si_6O_{18}$) which has a structure composed of ring silicate with a hexagonal structure arranged in parallel layers forming channels parallel to the *c*-axis. The layers are connected together by Al^{3+} ions located in octahedral coordination, while Be^{2+} ions are linked in tetrahedral coordination. Moreover, there are ionic substitutions generally occurring in beryl lattice sites such as Be^{2+} or Al^{3+} for Si^{4+} are Li^+ for Be^{2+} and Mg^{2+}, Fe^{2+}, Fe^{3+}, Mn^{2+}, Cr^{3+}, V^{3+}, and Ti^{4+} for Al^{3+}. In addition, other alkali ions (such as Na^+, K^+, Rb^+, and Cs^+) can accommodate inside the channel cavities to attain a balance charge, and water molecules can accommodate inside the hollow channels within the silica tetrahedral rings. The green color of emerald is from trace concentrations of chromium (Cr^{3+}) and/or vanadium (V^{3+}) by substituting small amounts in the aluminium (Al^{3+}) position [30].

FTIR spectroscopy as a vibrational molecular spectroscopic technique has been widely applied to identify beryl, especially emerald. The characteristic vibrational modes of Be–O and Si–O, Al–O, and Al–O–Si in emerald structure are observed in FTIR spectra [72]. Main IR absorption bands of emerald are summarized in table 3.3. Moreover, the bands due to hydroxyl groups associated with lattice are also noticed. Most of FTIR applications for emerald analysis are the differentiation between natural and synthetic emeralds [71, 73–75], identification of filler substances in emerald [76], and origin identification.

Table 3.3. FTIR peak assignments of emerald and treated emerald.

Peak position (cm^{-1})	Peak assignment
460–470	Si–O–Si bending
490–530	Si–O–Al bending
950	Si–O bending
1010–1050	Internal Si–O stretching vibration
1170–1200	Be–O vibration and Si–O stretching vibration
1200–1210	Si–O stretching vibration
3600–3640	Type II water
3595	Stretching vibration involving H_2O–Na–OH_2
3655	OH group bonded to alkali ions
3670–3650	Type II water
2746, 2815, 2887, 2625, 2970, shoulder at 2917 cm^{-1}	Chlorine-bearing vibration
3100–2800	C–H stretching; indicator of organic filler in treated emerald
1700–1600	C=C, C=O stretching; indicator of organic filler in treated emerald

3.5.1 Identification of natural and synthetic emeralds

In general, gemmologists recognize the characteristic inclusion of natural and synthetic emeralds during microscopic inspection for their separation and identification. Besides, FTIR spectroscopy is a powerful chemical analysis for differentiating natural and synthetic emeralds. Since the synthetic gemstone has similar chemical components to natural gemstone, in some cases, the absorption bands related to the main chemical structure in both materials cannot distinctly differentiate, especially low-quality spectrum. For emerald, thus, the characteristic of water absorption bands and some trace elements in its impurities utilizes for the discrimination of natural and synthetic emerald [75, 78, 79]. For flux-grown synthetic emerald (Chatham, Gilson, etc), the formation of flux emerald differs sharply from the natural emerald, notably by the lack of water. Therefore, the flux products usually are easily differentiated from natural and hydrothermal emerald based on the absorption of water molecules. That is because no water absorption bands are observable in the flux emerald's spectrum. On the contrary, all natural and hydrothermal synthetic emeralds contain water in the relatively large channels inherent to the structure of emeralds. Therefore, the distinction of natural and hydrothermal emerald must be distinguished by IR absorption of water molecules [73–75].

IR spectrum is used to characterize the water molecules in the beryl. The molecules were classified into two types, type I and type II, depending on their arrangement inside the channel cavities of beryl and the alkali content [77, 80]. Type I transition is water molecules in alkali-free beryls and has its twofold symmetry axis perpendicular to the crystallographic *c*-axis of the beryl. Type II is water molecules associated with nearby alkalis and has its twofold symmetry axis parallel to the crystallographic *c*-axis. In hydrothermal-grown synthetic emeralds, water molecules that are not bound to alkali ions (type I water molecules) are present, while type I and type II water molecules are found in natural emeralds. Hence, IR absorption of water molecules plays a helpful role for the discrimination between natural and synthetic emerald that occurs as type I and II in the natural emerald and as type I only in the hydrothermal emerald [73–75].

The peak position of type I and II H_2O (both natural and synthetic) has previously been observed at frequencies ranging from 3500 cm^{-1} to 4000 cm^{-1}. In particular, the molecular absorption band of type I H_2O is in the region of 3630–3600 cm^{-1}, while that of type II H_2O is in the region of 3670–3650 cm^{-1}. In summary, an absorption band around 3694 cm^{-1} is assigned to type I water molecules. The band at 3592 cm^{-1} and 3654 cm^{-1} are caused by the symmetric and asymmetric stretching mode of type II water molecules and/or hydroxyl groups, respectively. Most probably, the band at 3592 cm^{-1} is due to type IIa water molecules that are bound to alkalies (lithium and/or sodium) in the sequence H_2O–Na–OH_2. Another band at 3655 cm^{-1} is assigned to type IIb water molecules or hydroxyl groups that are associated to alkalies in the sequence H_2O–Na–€ or HO–Na–€, with € representing vacancies in channel sites of the beryl structure [78–83]. In addition, a weak absorption at 3230 cm^{-1} was assigned as an overtone of water molecules. The peak assignment of type of water in emeralds is summarized in table 3.3.

In addition, an absorption band related to chlorine bearing is another diagnostic range in the mid-IR which is a characteristic of some hydrothermal synthetic emerald. The synthetic chlorine-bearing sample revealed a series of three sharp bands between 2500 cm^{-1} and 3100 cm^{-1}. There are 2746 cm^{-1}, 2815 cm^{-1}, and 2887 cm^{-1}; two broad absorption bands at about 2625 cm^{-1} and 2970 cm^{-1}; and a smaller shoulder at 2917 cm^{-1}. The hydrothermal synthetic emeralds that revealed an identical series of absorption bands in all the chlorine-bearing products include Linde, Regency, and Biron. On the contrary, the chlorine-free synthetic emeralds do not show these series of absorption such as Russian and Lechleitner synthetic emeralds [73].

Moreover, Huong *et al* studied the natural and synthetic emeralds (hydrothermal and flux synthetic) from various sources by Raman and FTIR spectroscopy [71]. For FTIR measurement, the KBr pellet sampling technique was used for analysing different emerald samples, and good spectral quality was obtained. The results showed the absorption band at 1200 cm^{-1} (Si–O stretching), and its shoulder possibly corresponded to alkali content, which can be utilized for differentiating between natural and synthetic emeralds. An apparent maximum band of all synthetic samples exhibit at around 1200 cm^{-1} to 1207 cm^{-1}, while that of natural emeralds showed at about 1171 cm^{-1} to 1203 cm^{-1}. Moreover, some low-alkali natural and synthetic emerald samples reveal an overlap absorption band of the apparent maxima from 1200 cm^{-1} to 1203 cm^{-1}. All high-alkali emeralds appeared a distinct shoulder at about 1140 cm^{-1}, which related to alkali content. In case of low-alkali natural, this shoulder can be seen in Colombian emeralds (Chivor), but it does not appear in Nigerian samples. However, the shoulder was not observed in any of the synthetic samples [71].

Diffuse reflectance is a technique usually used with FT-IR instruments for gem analysis. However, it has limitations about the size of sample and the size of the sample cup of its accessory. Moreover, in the case of large sample size, diffuse reflectance spectra are almost found as the overabsorption band in the fundamental region and water absorption region (3500 cm^{-1} to 4000 cm^{-1}); the absorption is too strong to be distinguishable by using absorption of water and alkali content. Therefore, other FT-IR sampling techniques can be used for emerald analysis instead of the diffuse reflectance technique. As previously mentioned, various FT-IR sampling techniques were designed for the specific application of emerald analysis, and different sampling techniques encounter different difficulties. Although good spectral quality can be obtained by the KBr pellet technique, the method is not suitable for faceted emeralds because of the destructive nature of the sample preparation process. Therefore, the ATR sampling technique is an alternative sampling technique that is useful to emerald analysis. Examples of emerald spectra that were collected by diffused reflectance and ATR technique are shown in figure 3.8. As shown in the figure, the absorption band in the water and fingerprint region (2000–400 cm^{-1}) of the diffuse reflectance spectrum revealed overabsorption, while that of the ATR spectrum clearly presented a high quality of spectral feature.

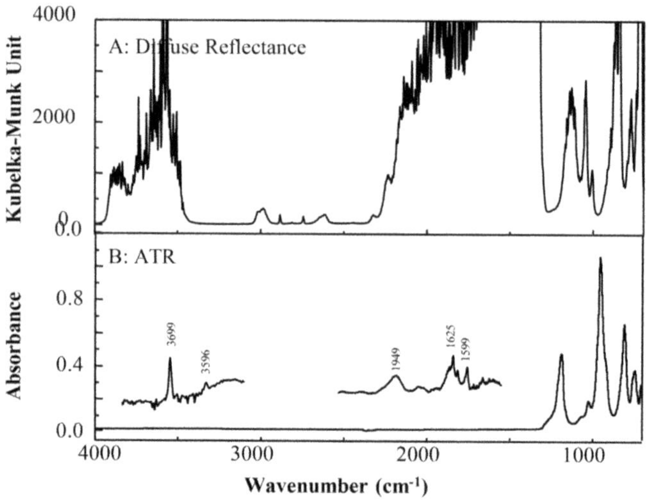

Figure 3.8. IR spectrum of natural emerald acquired by (a) diffuse reflectance and (b) ATR technique.

Because of the power of the ATR sampling technique, it was employed for identification of the water molecules in natural emeralds from various localities and comparison of hydrothermal synthetic emeralds [14]. Moreover, a peak finding of ATR spectra by curve fitting with Gaussian functions can resolve the overlapped bands in the water absorption and fundamental region. The curve fitting of OH (3800–3400 cm^{-1}) and Si–O (1000–1300 cm^{-1}) stretching bands was studied by Branca *et al* [84]. The result showed that the Gaussian decomposition method equipped with the second and fourth derivative of the spectra was possible to establish a relation between the position and the width of the subbands to the alkali content inside the beryl channels [84].

3.5.2 Origin determination

The provenance and origin of gemstones is one of the important factors for gem appraisal. Therefore, origin determination is an important task for gemmologists and buyesr. FTIR spectroscopy is one of the spectroscopic techniques utilized for origin identification of natural emeralds from different mining areas [77, 85, 86]. The results indicated IR spectra can be useful for determining the origin of emerald, and the mineralization environment type of emerald can be preliminarily identified from IR spectra. The obtained spectra, including mid- and near-IR regions, from different mine areas revealed different characteristics for emerald.

In addition, non-destructive and non-invasive methodologies such as EPMA and micro-FTIR spectroscopy were employed for the characterization of an emerald dating from the first century AD found in Oplontis (Torre Annunziata, Naples, Italy) [85]. The experimental data from both methodologies were also statistically treated in order to classify the emerald's orgin. All reported data and statistical analysis can support the hypothesis that the emeralds were traded to Rome from the

El Sikait deposit, one of several famous Egyptian mines. It can be implied that the FITR data equipped with statistical analysis can be useful for the identification of emerald origin.

Karampelas *et al* investigated gem-quality emeralds from the most important sources (Afghanistan, Brazil, Colombia, Ethiopia, Madagascar, Russia, Zambia, and Zimbabwe). Chemical and spectroscopic data using LA-ICP-MS, UV–Vis–NIR, and Raman and FTIR spectroscopy were evaluated for discrimination [86]. Most of the spectra were unpolarized (or partially oriented) and acquired on randomly oriented samples.

The results from LA-ICP-MS measurement showed that emeralds from Afghanistan, Brazil, Colombia, and Madagascar have a relatively lower lithium content (7Li <200 ppmw) compared to other places (7Li >250 ppmw). In addition, accurate origin information for these emeralds can obtain from alkali element contents as well as scandium, manganese, cobalt, nickel, zinc, and gallium. UV–Vis absorption of iron-related bands can help for the separation of emeralds from Colombia and Afghanistan. Raman spectroscopy can also separate medium to high alkalis from Ethiopia, Madagascar, Zambia, and Zimbabwe from the others. For FTIR results, IR spectra of emeralds from Madagascar, Zambia, and Zimbabwe showed intense type II water as well as in some samples from Afghanistan and Ethiopia [86]. It is noticeable that all spectroscopic techniques can help to separate emeralds from different geological origin. However, the separation of different origin using these methods should be undertaken with caution because it still has an overlap between some samples from different geological environments.

3.5.3 Identification of resin-filled emeralds

Most varieties of beryl (aquamarine, morganite) are generally quite clear or do not contain obvious inclusions. In contrast, natural emerald presents many inclusions which usually affected its clarity. The inclusions are crystal, internal cavities or negative crystal, healing fissure and open fissure, etc. Regardless of origin, cut emeralds are sold with some enhancement for clarity. It is a general trade practice to oil emeralds, with a view to solidify their strength and improve appearance. To improve the clarity of emerald, the open fissures may be reduced by impregnating the gems into fillers or some permanent substances. Effective fillers must have refractive properties close to those of emerald. The common fillers used in emerald enhancement are oil, cedarwood oil, epoxy resin, Opticon, Palma (Palm oil), etc. The various colorless oils, either natural or synthetic, appear to be accepted in the trade. In some cases, green fillers are also applied to enhance the color of emerald including green Optican, Joban oil, etc. However, colorless filler applied in an emerald fracture is more difficult to detect than color substance [76, 87]. Thus, the identification of the fissure filler substances in emerald is required.

FTIR spectroscopy is a powerful non-destructive technique that is utilized for identification of the fracture filler in emerald. For the mid-IR region, untreated emerald from any natural source shows absorption in fundamental region wave-number region (below 2400 cm^{-1}) and OH-stretching region (above 3400 cm^{-1}). When an FTIR spectroscopy is performed for a treated emerald analysis, the obtained spectrum reveals a combination of the filler substance and the emerald spectrum. The FTIR spectrum of filler in emerald presents its characteristic band in two absorption areas, which can be used for identification. The first absorption region is between 1700 cm^{-1} and 1600 cm^{-1}, which shows absorption of organic substances such as C=C, C=O, and cyclic aromatic molecules. However, the emerald spectrum must show overabsorption in this region, or the absorption of filler can be overlapped with characteristics of emerald absorption. Thus, this area cannot be investigated by the filler. The second characteristic region for filler substances is between 2800 cm^{-1} and 3100 cm^{-1}. In this area, the spectrum of treated emerald clearly presented absorption bands related to C–H vibration and originate from the fillers that are suitable for filler's analysis. Moreover, FTIR spectroscopy is possible to differentiate type of filler substances. The spectral results enable distinction between emeralds treated with oil, with artificial resin, Canada balsam, etc [76].

3.6 Quartz

Quartz is one of the most abundant minerals in the Earth's crust. The basic structural unit of quartz is SiO_4 tetrahedron. Natural quartz always has trace elements in its crystal lattice, such as iron, aluminum, hydrogen, lithium, sodium, and potassium. The type and content of some trace elements are important for the cause of color. Moreover, the cause of color of quartz has been correlated not only with the incorporation of impurity ions or trace elements in structural sites of the lattice but also with color center formation based on heating, ionization irradiation [88, 89]. The most abundant quartz in nature is colorless and transparent and known as rock crystal. In addition, there are many varieties of colored quartz, such as citrine (yellow), rose quartz (pink), amethyst (purple), prasiolite (green), etc. The most common varieties used in jewellery are amethyst and citrine.

Certainly, synthetic colored quartz can easily be produced in a laboratory by adding the appropriate color-producing elements to grow the crystal. Synthetic colored quartzes are widely used in jewellery because of their homogeneous colors compared with their natural counterparts. Colored varieties, especially amethyst, citrine, ametrine, prasiolite, and blue quartz, are synthesized in various countries such as Russia, Japan, and China. The identification of some types of synthetic quartz is straightforward with standard microscopy by the observation of color zoning, twinning, or inclusion [90]. However, the separation of natural from some synthetic quartzes is often no easy task, especially in cases of the highest quality products free of inclusions. Although natural and synthetic quartz have distinctly different crystal shapes, after faceting, most of the faceted quartzes can be nearly indistinguishable due to the high clarity and lack of inclusion.

3.6.1 Identification of natural and synthetic quartz

FTIR spectroscopy is one of the spectroscopic techniques for indicating color development and separating natural, synthetic, and treated quartz [91–97]. Moreover, it is possible to determine chemical change related to the change of color after irradiation [98, 99]. The mid-IR spectra of quartz is classified into four characteristic bands around 1000 cm^{-1}, 780 cm^{-1}, 695 cm^{-1}, and 450 cm^{-1}. The band in the 1080–1175 cm^{-1} region relates to Si–O asymmetrical stretching vibration, while that in 780–800 cm^{-1} region is Si–O symmetrical stretching vibration. The Si–O symmetrical bending vibration appears at 695 cm^{-1}, which is unique to the crystalline materials. The Si–O asymmetrical bending vibration reveals at 464 cm^{-1} [95].

Amethyst, prasiolite, and citrine are popular quartz widely used in jewellery and decoration. Amethyst is a purple variety of α-quartz whose color has been ascribed to iron impurities. The synthesized amethyst that is grown in either a near-neutral NH_4F solution can be identified by standard microscopy, while that grown in an alkaline K_2CO_3 solution cannot [91]. Thus, standard gemmological testing is not always sufficient to identify synthetic amethyst. LA-ICP-MS is an effective technique for separating natural and synthetic quartz by plotting Ti+Cr versus Ga [96]. Unfortunately, this technique is destructive and is too expensive for the identification of a common gemstone such as quartz. The technique is not widely available for routine gem analysis. According to Karampelas *et al*, FTIR spectroscopy with a high resolution (0.5 cm^{-1}) can provide a criterion for separating the synthetic and natural quartzes. The results reveal the FTIR spectrum has shown the band of OH stretching at 3595 cm^{-1} in the vast majority of natural amethyst [91]. The 3595 cm^{-1} band may be due to a boron-related defect in the amethyst lattice. Although the 3595 cm^{-1} band can occur in synthetic amethyst, it has a much larger FWHM value than that in natural specimens. If the absorption band of 3595 cm^{-1} on a quartz sample has a width of 3.3 ± 0.6 cm^{-1}, it can be concluded that the sample is natural. It can be concluded that the quartz sample is synthetic in case of there is no 3595 cm^{-1} absorption, or it has an FWHM of 7 cm^{-1} or more. The IR absorption criteria were only useable for amethyst. The 3595 cm^{-1} band may be due to a boron-related defect in the amethyst lattice [91, 92].

In the case of blue synthetic quartz, FTIR spectroscopy is one of the analysis techniques used for studying the cobalt-doped quartz crystal mechanism of synthetic blue quartz [97]. The synthetic cobalt-doped quartz was grown hydrothermally in steel autoclaves, and the as-grown quartz samples are originally yellow. The obtained sample acquired blue coloration after prolonged heat treatment times at 500 °C. The IR absorption spectra for the as-grown and heat-treated (500 °C) sample show three sharp bands localized at 3390 cm^{-1}, 3433 cm^{-1}, and 3580 cm^{-1}. The band at 3390 cm^{-1} was attributed to alkali-OH defects, which was observed in synthetic quartz. The bands at 3400 cm^{-1} and at 3580 cm^{-1} are due to OH^- perturbed by Li^+ and K^+ ions, respectively. After heat treatment, the earlier two sharp bands increase slightly in intensity, whereas the third band at 3580 cm^{-1} is not affected. Moreover, a new absorption peak appears at 3226 cm^{-1}, and its intensity increases continuously with the thermal treatment, which is probably due to the formation of Si–OH bonds and the lattice arrangement of the quartz lattice. In addition, both natural and synthetic quartz revealed the same broad band at 3230 cm^{-1} that corresponds to $OH–H^+$ vibration band. Besides, this band is also presented in heat-treated amethyst and high temperature boric oxide glass [98].

3.6.2 Characterization of heat treatment and irradiation

FTIR spectroscopy can be employed for characterizing heat treatment and irradiation of quartz from different localities [99]. The quartz samples were natural amethyst, citrine, and prasiolite, as-received, gamma-irradiated, UV-irradiated, and heat-treated conditions. It is noticeable that the identification of quartz must use a high resolution and high number of scans. FTIR spectra were determined in the spectral range from 3700 to 2600 cm^{-1} using a resolution of 1 cm^{-1} and 128 scans. The results reported that the gamma radiation had a great effect on the color of amethyst and citrine. The light-colored quartzs can increase their color upon gamma radiation, and it was possible to fade their color by heat treatment. The obtained FTIR spectra revealed that broad bands centered at about 3200 cm^{-1} and 3300 cm^{-1} were not affected by gamma radiation. The intensity of 3306 cm^{-1} and 3366 cm^{-1} was increased upon gamma irradiation and decreased upon heat treatment. In contrast, the intensity of absorption peak at 3349 cm^{-1}, 3393 cm^{-1}, 3436 cm^{-1}, and 3580 cm^{-1} decreased upon gamma irradiation. It may be because of the combination between alkali ions and hydroxyls. In addition, the low intensity band at 3545 cm^{-1} is slightly affected by gamma radiation and also has a slight influence on the color. For the prasiolite sample, the band at 3595 cm^{-1} related to B-related defects in quartz [99]. Karampelas *et al* reported that this band may be observed in the spectrum of natural quartz, and it is rarely observed in synthetic samples. However, it was noticed that this feature is slightly affected by gamma radiation, and it seems to have a slight effect on the color of quartz varieties [91, 92].

3.7 Jade

The term jade has been historically used to distinguish polycrystalline green minerals that were used as an emblem of prestige in different cultures from around the world. It is an ornamental mineral, mostly known for its green varieties, though it appears naturally in other colors as well, notably yellow and white. Jade can be classified into two different types based on the different silicate minerals inside them: nephrite and jadeite. Nephrite is a variety of calcium and magnesium in the amphibole group of minerals. The chemical formula of nephrite is $Ca_2(Mg,Fe)_5Si_8O_{22}(OH)_2$. Jadeite is a variety of sodium and aluminium in the pyroxene group of minerals with composition $NaAlSi_2O_6$. It is a polychromatic mineral due to numerous substitutions in its crystalline structure. Jadeite jade has a large variety of green shades, and it is also found in various colors such as white, black, violet or lavender, blue, yellow, orange, and red. Although jadeite and nephrite have distinctly different mineral compositions, they have very similar physical properties, especially visual appearance. Only trained observers or gemmologists with significant experience can reliably differentiate them without gemmological testing equipment [30, 100–102].

FTIR spectroscopy is a powerful technique for the characterization of jadeite and nephrite, and their peak assignments have been reported [100–102]. Lan *et al* studied jade using the KBr pellet technique and specular reflection technique [100]. The samples were black and green nephrite and jadeite jade. For the transmission technique, the technique is useful for differentiating black and green nephrites and jades because the spectra of both nephrites obviously showed a difference in peak position in the fundamental region (400–1600 cm^{-1}). The obtained spectra of black and green nephrite showed similar peak positions at 756 cm^{-1}, 687 cm^{-1}, 507 cm^{-1}, and 467 cm^{-1}, whereas those of jadeite were different. Since jadeite does not have a hydroxyl (OH) group in its structure, OH stretching vibrational band (around 3400–3600 cm^{-1}) does not occur in jadeite spectra. On the contrary, the nephrite spectra showed weak absorption bands of OH stretching at about 3672 cm^{-1}, 3658 cm^{-1}, and 3642 cm^{-1}. Peak positions of black nephrite jade and green nephrite jade are summarized by Lan *et al* [100].

Specular reflection technique is a popular method for the identification of jade [102–104]. The obtained reflectance spectrum is conversed to absorbance spectra by using the Kramer–Kronig transformation. The resulting spectra show slightly different appearances to those from the transmission technique. It is because the specular reflectance depends not only on chemical composition but also the reflectivity and transmittivity of the surface of materials [14]. However, the results showed that obtained reflectance spectra of different jades allowed to distinguish black nephrite jade from green nephrite jade and from jadeite jade.

3.7.1 Identification of jade enhancement

Due to jade's popularity and rising cost, many imitations and treated jade have appeared on the gems market. In addition, most jades on the market, both jadeite and nephrite varieties, have undergone enhancement to improve their color, luster, and stability. The treatments such as heat, wax, dye, bleach, acid, and polymer injection are widespread to enhance for low-quality jade. In trade, there are three main methods of enhancement, referred to as the ABC treatment system. The grading system explains the degree of enhancement to describe the level of treatment as the following:

Type A jade has not been treated in any way except surface waxing to improve its luster.

Type B jade has undergone intense treatment by chemical bleaches and/or acid and impregnating with a clear polymer resin. The jade is soaked in acid to remove the stains. The results show a significant improvement of color and transparency of jade. Currently, FTIR spectroscopy is the most accurate technique for identification of polymer in this jade.

Type C jade has been artificially stained or dyed. The effect of this treatment results in translucency and may show dull brown or concentrated dye.

In addition, some jade is treated with acid and dyed polymer, or composite jade, called type B+C and type D jade, respectively.

Type B+C jade is a combination of B and C, treated with acid and dyed polymer.

Type D jade refers to a composite stone such as a jade top with a plastic backing.

Although many of these treatments are part of the jade trade, any significant jade purchase has been confirmed to obtain a laboratory report to ensure the type's jade. That is because, in some cases, a portion of the coating of type B and C jade had peeled off and exposed the true color of the jadeite. In general, the characteristic of coated jadeite was determined using standard gemmological equipment, DiamondView imaging, microscopy, and spectroscopic techniques [105]. DiamondView can reveal a difference in fluorescence reaction between the surface coating and uncoating beneath the jadeite. Microscopy may not show enough detail to identify the treatment, but FTIR spectroscopy is an easy method to discriminate type of jade. The IR spectrum of coated jade revealed four characteristic bands at about 2856 cm^{-1}, 2873 cm^{-1}, 2928 cm^{-1}, and 2958 cm^{-1}, which indicated the presence of an organic polymer or coating material on jade's surface. These bands do not appear in the spectra of uncoated jadeite. The bands at 2873 cm^{-1} and 2928 cm^{-1} related to the symmetric and asymmetric stretching vibrations of CH_2, while the 2958 cm^{-1} band was caused by the asymmetric stretching vibration of CH_3. The results indicate the existence of coating material as epoxy resin [105–107]

3.8 Turquoise

The mineral turquoise is one of the oldest gemstones used in jewellery and decoration, dating back 70 centuries, to ancient Egypt. Nowadays, turquoise is popular in fine jewellery as well as in various cultures. It is common in the American Southwest, where most of its value is correlated to art, jewellery, and Native American culture [108–110]. Since the supply of high-quality turquoise is limited due to the porosity characteristic for massive aggregates, it causes the presence of a large number of treated turquoise and imitations on the gemstone market. The turquoise has been treated to various methods in order to improve its appearance and durability. The methods include impregnating with plastic, paraffin, and oil to enhance color or reduce porosity. For turquoise imitation, turquoise was one of the first gems to be imitated by synthetics like glass, plastic, and pressed turquoise powder bonded by resin [111, 112]. In the case of synthetic turquoise, at the present time, synthetic turquoise is produced by the Gilson method. Synthetic and imitation can be separated from natural turquoise by basic gemmological equipment via their physical and gemmological properties. Although some treatment can be detected by observation with a microscope or use of a hot point, a relatively new, proprietary turquoise treatment adds difficulty to differentiate from an untreated turquoise counterpart with the routine gemmological technique. The detection of these treatments is very important, especially for the use of non-destructive methods.

3.8.1 Characterization of turquoise

The chemical composition of turquoise is a hydrous copper aluminum phosphate belonging to a group of minerals (the turquoise group). The general formula for the turquoise group may be written as $A_{0-1}B_6(PO_4)_4(OH)_8{\cdot}4H_2O$. The most common constituents at the A position are Cu^{2+} and Fe^{2+} and the B position are Al^{3+} and Fe^{3+}. However, the A position of some rare mineral in the turquoise group can be Ca^{2+} or Zn^{2+}. The different concentrations of Cu, Fe, and Al produce a wide range of turquoise colors [108–110]. The color of turquoise ranges from blue through shades of green to yellowish gray (usually occurs as sky blue, bluish green, apple green, or greenish gray). Differences in the Cu and Fe ratio are responsible for the differences in color of the samples. Blue turquoise has Cu at the A position and Al at the B position, whereas green turquoise (chalcosiderite) largely contains Fe^{3+} at the B position.

In this topic, turquoise is a representative sample for an analysis of opaque gemstones by FTIR spectroscopy. Most published papers on turquoise are, however, usually based only on IR or Raman spectroscopy. In the case of FTIR analysis, the KBr pellet method is a traditional sampling technique for an opaque material [109, 110]. Nevertheless, it is not suitable for the gems because of the destructive nature of the sample preparation process. DRIFT was used for turquoise mineral characterization. Normally, this technique is commonly used for samples that can be ground into a fine powder and mixed in a powder matrix such as potassium bromide (KBr). In the case of turquoise, for example, the sample powder of turquoise, about 0.050 mg, was mixed with KBr and placed into a sample cup. The procedure used KBr as a reference [109]. Although the results show high spectral quality, the sample preparation is destructive and not suitable for a high price product. However, tentative assignment of the turquoise spectra was assigned by Cejka *et al* [109].

At present, diffuse reflectance has gained popularity as a sample technique applied for gems analysis. The sample measurement is only to place the gem sample on the sampling cup. Although the diffuse reflectance technique is a highly effective non-destructive process approach for gems identification, it seems unsuitable for bead and cabochon shape. The spectral envelop of the opaque gemstone is clearly different from that of the corresponding transmission spectrum, as shown in figure 3.9. In some cases, at the low wavenumber region, the resultant spectrum exhibits derivative band shape arising from the specular reflectance component of the partial smooth surface in which the radiation does not penetrate the sample. The derivative band shape becomes a particular problem, and it is difficult to verify a certain peak position for chemical composition interpretation.

The turquoise spectrum shows characteristic peaks in the region below 900 cm^{-1}, which is mainly absorption due to coupled motions of the frameworks of tetrahedra and octahedra. The P–O stretching vibrations of the phosphate group $(PO_4)^{3-}$ are located between 1200 and 900 cm^{-1}. The bands at 4000–2500 cm^{-1} are due to OH stretching vibrations of the OH groups, and the broad band at 1600 cm^{-1} is the H–O–H bending vibrations of the H_2O molecules. In the case of chalcosiderite, the

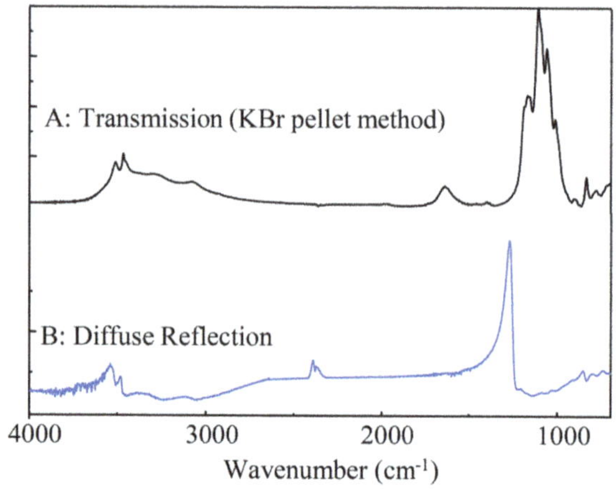

Figure 3.9. IR spectra of natural turquoise acquired by (a) transmission by KBr pellet method and (b) diffuse reflection.

OH stretching bands become broader with increasing Fe content, and the bands shift to slightly lower frequencies [108–111].

Some samples show the absorption spectrum of Si–O and C–O due to the presence of kaolinite and Ca carbonate, respectively, admixed with turquoise [110]. The characteristic OH stretching bands of kaolinite-group minerals can be observed between 3698 cm^{-1} and 3622 cm^{-1}, including very weak bands at around 3675 cm^{-1}, 3668 cm^{-1}, 3655 cm^{-1}, and 3648 cm^{-1}. The Si–O stretching bands of kaolinite occur at ~1000–1100 cm^{-1}, and the Si–O–Si bending band shows at ~470 cm^{-1}, as well as the OH bending of the inner OH groups at ~912 cm^{-1} [108–110]. In addition, the weak bands at 1430 cm^{-1} and 1460 cm^{-1}, characteristic of the C–O stretching vibrations of the carbonate group $(CO_3)^{2-}$, can be also observed. Band component analysis via the Peakfit software package was undertaken to identify corresponding peaks in the spectrum [110].

3.8.2 Identification of treated and imitated turquoise

Fritsch *et al* had studied the identification of natural and Zachery-treated turquoise by using various standard gemmological and spectroscopic techniques [112]. Although the turquoise shows a slightly unnatural color and blue color concentrations in and around fractures, the experimental results reveal it cannot be detected by standard gemmological techniques. The article used reflectance IR spectroscopy and EDXRF to analyse the samples. The reflectance IR spectra of natural and treated turquoise reveal similar spectral features which represent the vibration of PO at 1125 cm^{-1}, 1050 cm^{-1}, and 1000 cm^{-1}. It is not a significant difference. In this article, the effective method for differentiating untreated and Zachery-treated turquoise is EDXRF. The chemical analysis revealed that Zachery-treated turquoise has significantly more potassium than that of an untreated sample. In addition, DRIFT spectroscopy was used for the characterization of polymer-impregnated turquoise. The result spectrum shows absorption bands at 1725 cm^{-1}, 1734 cm^{-1}, and 1744 cm^{-1}, which related to polymers used for stabilizing turquoise [112]. These bands do not occur in untreated turquoise or natural turquoise.

For other types of turquoise, gem-quality turquoise with various colors, such as chalcosiderite (Fe-rich) and faustite (Zn-rich), were investigated by the reflectance method; their IR peaks assignment have been summarized [108]. The results show that Cu-rich blue turquoise clearly has a lower reflectivity than Fe-rich chalcosiderite and Zn-rich faustite, especially in the phosphate group $(PO_4)^{3-}$ spectral region. In the case of high Fe content as chalcosiderite, the peak resolution of P–O stretching vibration below 700 cm^{-1} was reduced, and absorption peaks at 609 cm^{-1} and 648 cm^{-1} loses which distinguishes from turquoise. In faustite, the absorption peak at 1057 cm^{-1} separated into two peaks at 1061 cm^{-1} and 1047 cm^{-1}. Moreover, the reflectivity ratio of the double peaks at 781 cm^{-1} and 833 cm^{-1} representing OH is negatively correlated with Cu content. This is utilized to identify the three turquoise categories [108].

Moreover, FTIR spectroscopy can be used for studying historical turquoise glass beads. The opaque turquoise glass beads from museum collections of Russia were identified using micro-FTIR spectroscopy via ATR mode. The visual appearances of the beads look like turquoise, but their chemical composition is totally different from the turquoise mineral. The main spectra revealed a strong and wide peak at 1100 cm^{-1} assigned to Si–O stretching and a weak wide peak at 800 cm^{-1} assigned to the Si–O bending vibrations [113].

3.9 Application of machine learning algorithm to gemstone classification

Generally, gemstones are minerals or rocks occurring in preferential sites on the Earth. They are produced from the weathering of minerals for such a long period of time to form charming crystallization in different color shades. Due to the properties of durability, hardness, and toughness of gemstones, they relate to the ability of the gems to preserve their color appearances without loss of color dissolution by any chemical and environmental factors (e.g., temperature and humidity) over a period of time. The value of gemstones is directly related to durability, rarity, demand, traditional impact, portability, and physical beauty (e.g., its color, luster, cutting, etc). Gemstones and jewellery are currently some of the important products contributing to a large part of trading economics in the world market, especially in Asia and Southeast Asia countries. This involves the circulated export value of more than 10 billion USD per year. In the trading market, only a few that actually meet the standards are exceedingly rare and precisely expensive. Therefore, determining the quality of gemstones is incredibly important to reassure both jewellers and customers.

Unfortunately, there are only a few standard classification protocols available for a specific type of gemstone such as the diamond. The Gemological Institute of America offers a fixed scale to grade and classify diamonds based on their qualities, while there is no international standard for grading the quality in colored gemstones (e.g., sapphire, emerald, alexandrite, tanzanite, etc). In recent years, the small percentage of high-quality gem material has been produced from a given mining locality, while a much larger percentage of lower quality material (with poorer color and/or unsatisfied clarity) is recovered that has little market value. Unfortunately, nowadays, chemical and physical treatments such as temperature, pressure, irradiation, diffusion of mineral, and pigment insertion are used to improve the color quality of gemstones. They have been claimed as the 'synthetic gemstone' or 'treated gemstone' consisting of similar chemical compositions and its appearance as the natural gemstone. These synthetic treatments have been continuously developed to transform the low quality of gem materials in order to enhance their presence, including inclusion, color, and clarity, and increase their commercial values. Therefore, it is extremely difficult to differentiate natural gemstones from synthetic ones, even for a gem speialist. This enables the promotion of synthetic gemstones by taking advantage of the inefficiencies of standard traditional or sophisticated modern identification methods which require expert supervision and are time consuming and extremely expensive. Moreover, it is not guaranteed that the setup protocol of considering only the physical appearances (e.g., color and inclusion) can effectively reveal the quality of gemstone. With a lack of standard index, one of the trickiest things in the world of gem markets is gemstone recognition and identification.

As already discussed, most precious gemstones are minerals consisting of different chemical formula and chemical elements. Minerals usually form as a result of inorganic processes of the chemical elements (e.g., $Ca_2Al_3(SiO_4)(Si_2O_7)O(OH)$+Cr, Sr for tanzanite, α-Al_2O_3+Fe,Ti,Cr,Vr,Mn for sapphire) which occupy a unique internal atomic structure and finally crystallize with a small amount of transition metal. Therefore, the different types of gemstone should provide unique chemical contents and chemical functional bonds as they compose the different crystal structures at the atomic level. In this case, spectroscopic techniques have been used as powerful tools for mineral identification in gemmological laboratories around the world for the determination of gem compositions and for solving crystal-chemical problems in mineral structures. Importantly, spectroscopy is a non-invasive technique which could preserve the original gemstone in safe conditions without defects during the measurement. Elemental analysis is also important in gemology for identification, classification, and characterization of both natural and treated gemstones. Most of the chemical elements from both natural and treated gemstones are similar. Only trace levels of metals, e.g., alkali and alkali earth elements, especially the transition metals, assist to establish the geographic origin and environmental condition during the formation of gemstones. Moreover, the functional groups of the inorganic compounds create a unique pattern which could be easily monitored by spectroscopic techniques. Besides, spectroscopic technique requires simple sample preparation, non-destructive measurement, portability, and fast results, which makes it potentially useful as a new approach for gem identification. However, the protocol of using spectroscopic data involves multivariate analysis, which involve a load of variables (e.g., chemical contents and several thousands of wavelengths from vibrational spectroscopic data). It was not simple to study or interpret the huge amounts of data generated directly. The information is extremely complex for humans to manage; therefore, the statistical and mathematical techniques must be addressed.

One of the keys to successfully meeting the classification rules to discriminate the class of gemstone is to develop a new protocol of using spectroscopic data integrated with statistics, computation, and mathematics to provide in-depth and broad-range analysis. With the drastic development of computer hardware and software including the high-performance spectroscopic instruments, it offers such a great opportunity to originate a new platform for gemstone recognition and identification. This breakthrough technology came in the beginning of the 21st century by the availability of ultrafast personal and clustering computers, which not only give fast calculation but also provide enormous storage space with easy accessibility. Therefore, data treatment and data interpretation using machine learning on either elemental analysis or spectroscopic data could be automatically evaluated by the artificial intelligence (AI) model in a simple computer or mobile phone to perform rapid and fast determination with minimized bias from human decision. Despite the developments, the application of the machine learning model on gemstone spectroscopic data is increasingly mentioned in the related literatures as shown in figure 3.10. It reveals the evolution of the publication devoted to the application of spectroscopy on gemstones combined with machine learning since 2000–2020 following the keyword search of 'gemstone,' 'spectroscopy,' and 'machine learning.'

So far, there is only a small number of publications on the implementation of spectroscopy on gemstones, while the application of spectroscopic technique on diamonds has >10000 published works in the same period of time. It might depend on the valuation on type of gemstone. In the case of diamond, the value of diamond

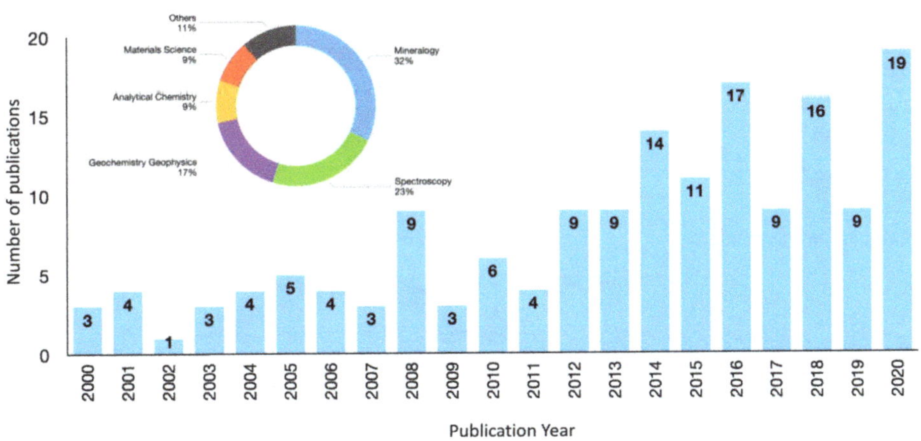

Figure 3.10. Bar graph shows an evolution of the publications devoted to the combination of spectroscopy and gemstone with the circle graph demonstrating the literature sources from 2000 to 2020 using keywords 'gemstone' and 'spectroscopy.' The results were searched from ISI Web of Knowledge (Web of Science)—Thomson Reuters.

is recognized worldwide in the trade markets. Diamonds can be claimed as the stable assets which preserve the worth price over the period of time. Therefore, it is not surprising that there are enormous studies of diamond classification using spectroscopic techniques. On the other hand, the color gemstone was unequally valued based on religions, beliefs, and cultures. For example, some color gemstones, e.g., jade and ruby, are preserved with very high value in Asian countries, but they have no value in European countries. However, this bar graph reflects an increase in the number of publications based on the application of spectroscopy on gemstones in major research themes such as mineralogy, spectroscopy, geochemistry, geophysics, analytical chemistry, and materials science. From the literature search, there is no doubt that spectroscopy is a predominantly important tool in the present day to achieve the aforementioned developments in the classification platform of the gemstones. For being an indirect way of saving time and acquiring more accuracy, the automated analysis was also implemented using machine learning calculation. A significant improvement of the platform can also be obtained by data analysis and visualization of large datasets to simplify and assist in reaching to the concise conclusions.

The general platform of gemstone identification using spectroscopic techniques combined with multivariate data analysis is summarized in figure 3.11. Particularly, the gemstone samples were collected, and then the spectra of the samples were obtained by advanced instruments in a standard research laboratory. Then, the acquired spectral data are exported into a computer and suitable software to either extract information or develop a statistical model. This protocol involves built-in intensive programming for very complex data; however, commercial software can also often be used. After validating the model, detection procedure and developed software can be organized and installed into the instruments for real-time analysis,

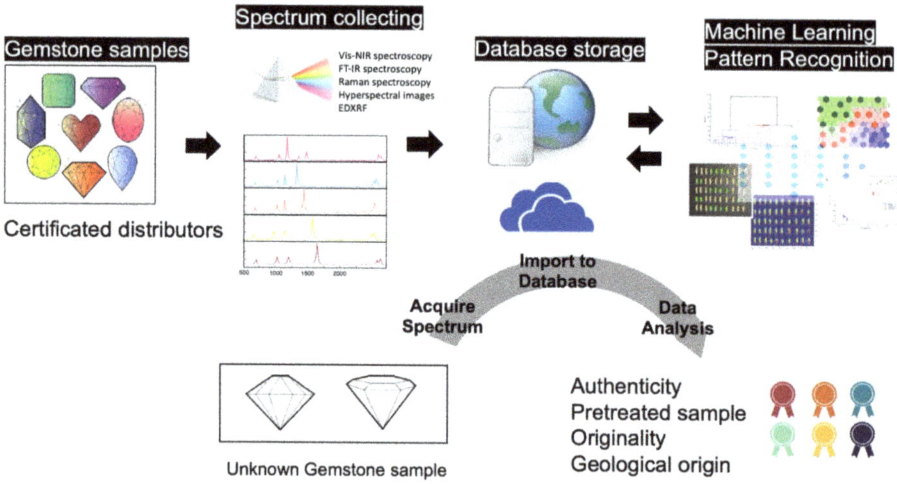

Figure 3.11. General platform of gemstone identification, classification, and inspection using spectroscopic data combined with machine learning analysis.

on-site analysis, and automated analysis. The platform could be divided into several steps as follows:

Gemstone sample collection: A set of gemstone samples should be collected from either the trusted source or the certificated distributors. Importantly, these collected gemstones are guaranteed as genuine or even treated ones. These gemstones can be used as the reference samples which are genuine and represent their authenticities and originalities. In each gemstone sample, the information of geological origin, grade, purity, clarity, physical appearances (e.g., color, clear index), treating process, and gem type should also be recorded. It should be noted that the higher the number of reference samples, the higher the accuracy of gemstone classification. To collect gemstones, experimental design which depends on the objective for gemstone classification should be carefully considered. For example, to classify the geological origins of the gemstone, the samples with the same type from different mining areas should be collected. In other cases, to identify nature gemstones out of the treated ones, samples of nature gemstones and synthetic gemstones should be gathered in order to create the protocol for differentiation.

Spectral acquisition: The spectral patterns of gemstone could be acquired from various spectroscopic techniques such as Vis–NIR, FT-IR, Raman, EDXRF, and Hyperspectral camera. The chemical information in gemstones will be revealed from different spectrometers. The electronic state of the inorganic complex presented in gemstones could be easily revealed by a Vis–NIR spectrometer. Generally, the color in gemstones should provide the distinctive absorption bands in the Vis–NIR spectrum. Color on gemstones is directly related to the individual maximum absorption peak in the spectrum. In the case of FT-IR and Raman, the vibration frequency of the chemical bond existing in the gemstones will be discovered. The information is crucial, as it directly conveys to the functional groups of chemical complex. The spectra create a unique fingerprint pattern to each gemstone sample. For EDXRF, it reveals the proportion of chemical elements presented in gemstones, especially the transition metals which are important for gemstone classification.

Database storage: The spectra information of gemstones is then imported to the database storage, which could be in the form of a server, personal computer, or storage cloud. It should be noted that the capacity of personal computers to store enormous volumes of spectra data is limited, and there is a limitation of accessibility. Nowadays, there is an enormous amount of research and scientific data that can be stored on internet and cloud systems, and sometimes they are free to access.

Data analysis: The data analysis software was used on the database to generate the characteristic spectral pattern corresponding to the gemstone. Machine learning calculations can be implemented on available programs such as MATLAB (www.mathworks.com), R-studio (www.rstudio.com), Google, etc. These might be stand-alone programs which can be installed and used for each personal computer separately and also web-based programs which are available to access via the internet. Most multivariate models are based on the defining classification model which is valid at the time, but it needs to be modified as time passes and more samples are collected. This encourages the field of machine learning to grow out of efforts to develop a toolbox of statistical and computer application for processing, analysing, and visualizing the available big data, which is sometimes called '*Big Data Analysis.*' Some sort of statistical and mathematical technique such as AI algorithm, which can update and adapt itself (self-learning approximation) for the better output without any external input, is needed in order to extract the essence of the data. This involves intensive programming and calculations combined with a searching algorithm in a cloud system. The specific spectral regions could be identified as the markers for gemstone classification. To generate the robust statistical model, an adequate number of spectra and gemstone samples should be used. After regenerating and validating the model several times, the model should be ready to be used for gemstone identification.

Classification of unknown gemstone: An unknown gemstone is transferred to the laboratory for inspection. The spectral data of the unknown sample will be acquired and then exported to the database on a computer, server, or storage cloud. The robust statistical model built from the reference gemstone is used to discriminate and classify the unknown to give certified outputs on authenticity, originality, and geological origin. Furthermore, the spectral data of the unknown can make it possible to update the statistical model to be further used for a next unknown gemstone. It could be seen that the model would more accurate when the process is continuously used.

Generally, the spectral measurement of gemstones can be obtained by various spectroscopic techniques. These can be defined as a characteristic pattern reflecting the chemical compositions of the gemstones, patterns, and fingerprints. The good quality of spectral data should be obtained from the high-resolution spectrometers. However, the data quality might be adequate by using only the standard spectrometers. Therefore, to gain the high accurate gemstone classification, one of the key factors to successfully meeting the standard is to develop the robust and powerful software for analysing and interpreting the spectral data in the simple way without using trail-and-error approaches. Data analysis based on machine learning helps here to extract the significant information with a smaller number of spectra. The methodology usually is based on experimental design, signal pre-processing, data exploring, optimization, calibration, pattern recognitions, and an artificial neuron network. Figure 3.12 shows an overview of applying data analysis throughout the flow of information from the beginning to the end of the spectral analysis process. Before starting the experiment, the strategy of experimental design should be considered in order to maximize the quality information with the smallest number of samples. This enables the evaluation of a large number of factors in the relatively small number of experiments using satisfied statistic and mathematic approaches. It minimizes the cost of operation without decreasing information quality and promisingly quantifying the model. After the acquired spectral data of the gemstones is generated, the treatment of acquired analytical spectra often called

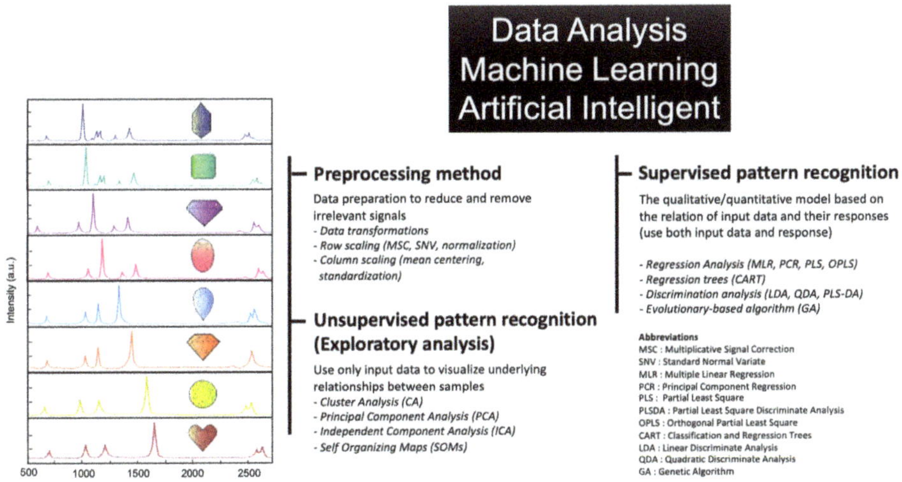

Figure 3.12. An overview of applying data analysis and machine learning protocol throughout the flow of information from the beginning to the end of the gemstone classification and identification process.

'*pre-processing*' are firstly introduced. This pre-processing method should be considered as an important step of the overall strategy of data analysis on gemstone classification. This step of data preparation can reduce and remove an undesired perturbation in the signals (such as light scattering effects) by signal denoising, data cleaning, outliner detection, transformation, scaling, and normalization. Next, the pre-processed data is projected to the clustering model which is often called '*unsupervised pattern recognition*' or '*exploratory analysis*' to visualize the main patterns or groups of samples in the data. From the estimation, the underlying relationships between gemstone samples is visualized with no further information such as response, class information, or concentration required. After the underlying information of the data was revealed, '*supervised pattern recognition*' was then performed by expressing the relationship between samples and the corresponding responses. The responses can be recorded in terms of class members associated with gemstone types, geological origins, authenticity, and pre-treated processes. The supervised pattern recognition is involved here to find the discrimination models used in a variety of applications such as prediction, calibration, classification, and regression. In the last step, the feature selection of spectral regions corresponding to the characteristics of chemicals in gemstones was performed. This involves both unsupervised and supervised task to seek the specific spectral regions which act as the class maker to differentiate the types of the gemstone. These obtained feature regions are crucial to be further investigated to monitor which chemical contents or chemical structures exist in the uniqueness of the gemstones. The detail of data analysis in each step will be discussed in the next section. In this chapter, we will summarize the generic methods used in data analysis but will not include the details of mathematical calculation here. They could be found elsewhere in the literatures and books [114–118].

Pre-processing methods: After spectral acquisition, pre-processing methods should be considered as an important step to treat the data prior to performing data analysis. The acquired data from spectroscopic measurements must be pre-treated before applying further manipulation to obtain true underlying information. In fact, it is not possible to collect spectra from a sample on the same environments and analyse those samples using the same instrument under identical experimental conditions. Variations due to collection date, light scattering effect, and the stability of the instrument and detection noise such as interference of chemical factors or errors in experimental process are presented in different samples in terms of peak height, peak shifted, and baseline, etc. These lead to the decrease in signal to noise ratio, and the mentioned factors may distort the important information extracted from the acquired spectral dataset. Fortunately, the unwanted variations from the raw data could be removed by using mathematical transformations. Different transformation methods have been developed in order to eliminate different artefacts or to adjust the data to be more linear. This further allows the mathematical model to calculate only on the relevant information, hereby increasing the rate of prediction performance and improving classification. Using an appropriate pre-processing strategy is crucial for data analysis or modelling. Using the inappropriate strategy might severely distort and harm the data analysis process. Specific spectroscopic artefacts may not be completely removed by certain pre-processing. Pre-processing methods can also provide different bad side effects, as it possibly can remove essential information or make unimportant information more impactful. Due to these reasons, these further complicate the selection of good pre-processing methods, as they can also have different interactions with each other. In practice, the 'global' or 'perfect' pre-processing strategy may therefore not even exist. The data could be in a form of matrix in which row is gemstone samples and column is independent variables such as wavelength, wavenumber, Raman shift, or chemical element. There are three principal ways to scale the data matrix. In the measurement, the acquired spectra might be influenced by the non-linearities introduced in the experiments, especially the spectroscopic measurement on the gemstones. The acquired spectra from spectroscopic techniques can be strongly influenced by a light scatter of inflated surfaces of the samples or the inclusion on the gemstone surface. It is not possible to control this influence to be equal in every sample. Therefore, it will affect the signal, which leads to a background shift from undesired light scatter effects during the measurement. Analysing the absolute intensity of each sample will not suitably reveal the real relationships between the groups of samples. In this case, baseline correction combined with normalization are routinely used to minimize the baseline shift from the scattering effects. The goal of this pre-process step is to improve a subsequent data analysis such as exploratory analysis, calibration model, and classification model. A general scheme of pre-processing spectral data prior to performing data analysis and machine learning protocol and the influences of pre-processing methods including baseline correction, normalization, and column scaling on the original spectral data acquired from gemstones is summarized in figure 3.13.

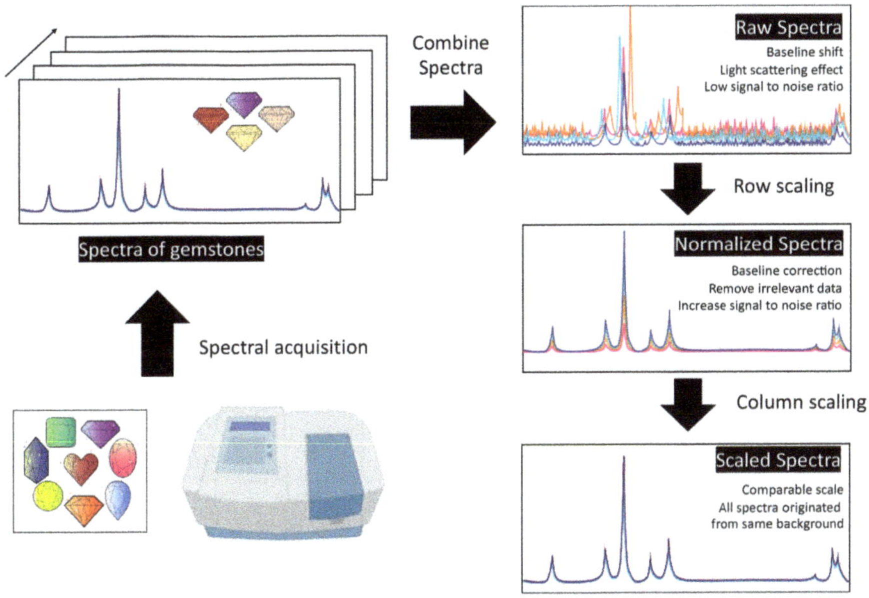

Figure 3.13. A general scheme of pre-processing spectral data prior to perform data analysis and machine learning protocol.

Unsupervised pattern recognition: After the appropriate pre-processing methods have been applied, the pre-processed data matrix contains the relevant information as a majority with a small proportion of irrelevant artefacts. It is also in comparable scales and is ready for the further data analysis. Nowadays, instruments, data storage methods, and computers are dramatically developed with high resolution. They provide a lot of information even for just only a single sample. Therefore, the spectroscopic datasets are getting largef and more complex. It is impossible to interpret and visualize the underlying relationship of samples directly from these data by human routine analysis. Hence, some techniques are needed to organize the information in easier forms for interpretation with no further knowledge such as class information, concentration, etc, required. Pattern recognition is a process of description, grouping, and classification of underlying patterns related to the type of samples. An unsupervised approach can be split into two basic principles involving dimension reduction and cluster/group analysis. For data dimension reduction, this is fundamentally based on a multivariate linear transformation in order to decompose the original data which involves several variables and projects them into the new dimension space involving only a couple of correlated variables (latent variables). The samples in the new dimension space could be easily visualized using 1D, 2D, and 3D plots to monitor the correlation of sample based on their spectroscopic data. In another case, cluster analysis is an algorithm to discover the characteristic patterns on the spectroscopic data and then cluster those samples with similar characters together, at a farther distance from the others. Unsupervised methods are generally thought of as 'data exploratory driven' because the defined

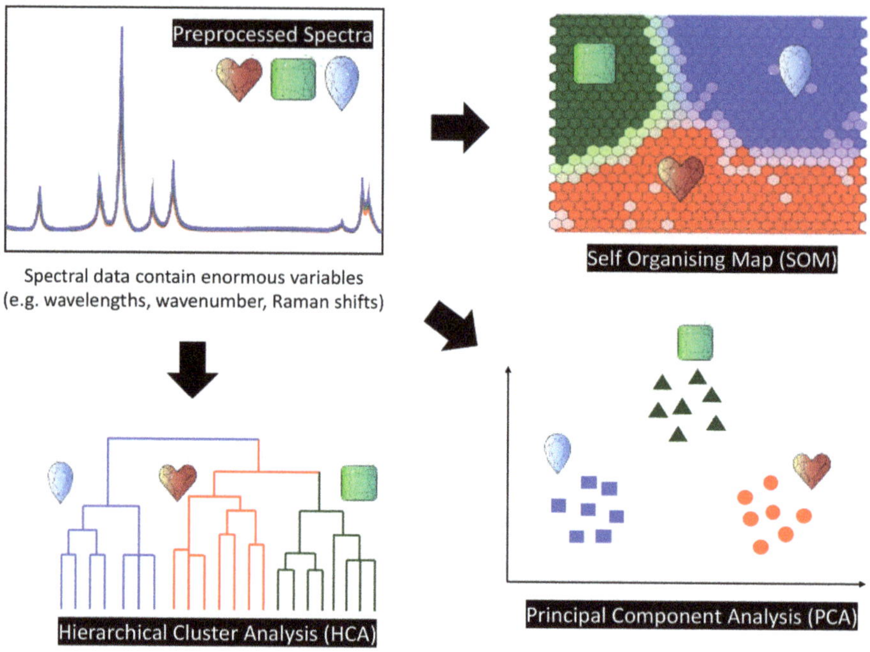

Figure 3.14. Under the unsupervised pattern recognition approach, the spectral data projected to the new dimension space to visualize the cluster of sample groups using HCA with Euclidean distance and nearest linkage, graphical representation of samples in PC scores from PCA, and graphical representation of the SOM map of each gemstone sample using the SOM algorithm.

clusters are based solely on the data used for further modelling. In this case, the ability and affinity of the model to make predictions and classifications about unknown samples is not considered. The well-known algorithms of unsupervised pattern recognitions are hierarchical cluster analysis (HCA), principal component analysis (PCA), principal coordinate analysis, and self-organizing maps (SOMs). In addition to these methods, several others such as independent component analysis were developed based on unsupervised principles that are fully described elsewhere [106–108, 111–113]. The example of unsupervised pattern recognitions HCA, PCA, and SOMs on the pre-processed spectral data of gemstones (three types) are illustrated in figure 3.14. The data in the form of spectra can be projected in the new dimension space to visualize and interpret the relationship and clusters of samples.

Supervised pattern recognition: Supervised pattern recognition is mainly used for estimation, prediction, and classification. Prediction and classification problems are the most generic problems in pattern recognition tasks. The basic concept is to find the protocol that will separate pre-defined groups of gemstone samples, which is often called 'classifier,' and then apply the created rule to determine which group unknown samples belong to. To generate the classifier model, the relationship between samples and responses (they can be class, group, pre-treated process, etc) are revealed by measuring a series of samples and then by building the satisfied mathematical model. Once the model is generated, it can be used to either predict the response or classify the group of unknown samples. These are called 'supervised' pattern recognition tasks, which therefore differ from unsupervised methods because the main aim is classification and prediction rather than exploration. It can be seen that the decision rule which represents a boundary line between the classes is defined by the classifier, and there are several different types. In fact, there are an enormous number of methods available in the literature; however, the majority of methods can be divided into two main areas, which are those based on statistical models and those based on machine learning techniques. The statistical classifiers [114–116] are Euclidean distance, linear discriminate analysis (LDA), and quadratic discriminant analysis (QDA), which involve indicators like the average and covariance to model the pre-defined groups and then assign the class membership of unknown samples based on the smallest distance from the centroid of the pre-defined classes. Other methods which involve the higher level of statistical and mathematical calculation are based on regression techniques [114–116] of the latent variable model such as multiple linear regression, principal component regression, and partial least square regression. Recently, the complicated machine learning approaches have been developed to deal with complex non-linear problems. The simplest method is the *k*-nearest neighbours classifier. A major area of research in machine learning is neural networks, which are algorithms based on the function of biological neurons. There are several types of neural networks depending on the connections linking the different layers existing in the network. Examples of feedforward neural networks are linear perceptron, multilayer perceptron, radial basis function networks, probabilistic neural networks and learning vector quantization [122], supporting vector machines [123], and supervised SOMs [119–121]. Examples of the linear, non-linear, and neural network classifier of classes of gemstone is generated using LDA, QDA, and supervised SOMs, respectively, and are illustrated in figure 3.15. This includes how to apply the decision boundary from the classifier to determine a group of an unknown gemstone based on its spectral data.

Spectral feature selection: One of the most important tasks in pattern recognition is to determine which variables are the best discriminators. However, the spectral data is continuous; therefore, to select the discrete variable (e.g., wavelength, wavenumber, Raman shift) is not sensible here. Feature selection will be appropriate to be used in this case. The feature selection appears in different areas such as pattern recognition, machine learning, data mining, and multivariate statistical analysis. The aim of feature selection is to select those spectral regions that allow the improvement of the gemstone classification output. It should be noted that most of the spectral region is irrelevant (especially the region with no intense peak) for the main objective of the study. They may originate from the instrument background or factors that are irrelevant to the problem in hand, so it is necessary to search for a small combination of spectral regions that are required for a specific factor or classifier. In fact, it is impractical to investigate all possible combinations of features because there are a large number of possible combinations.

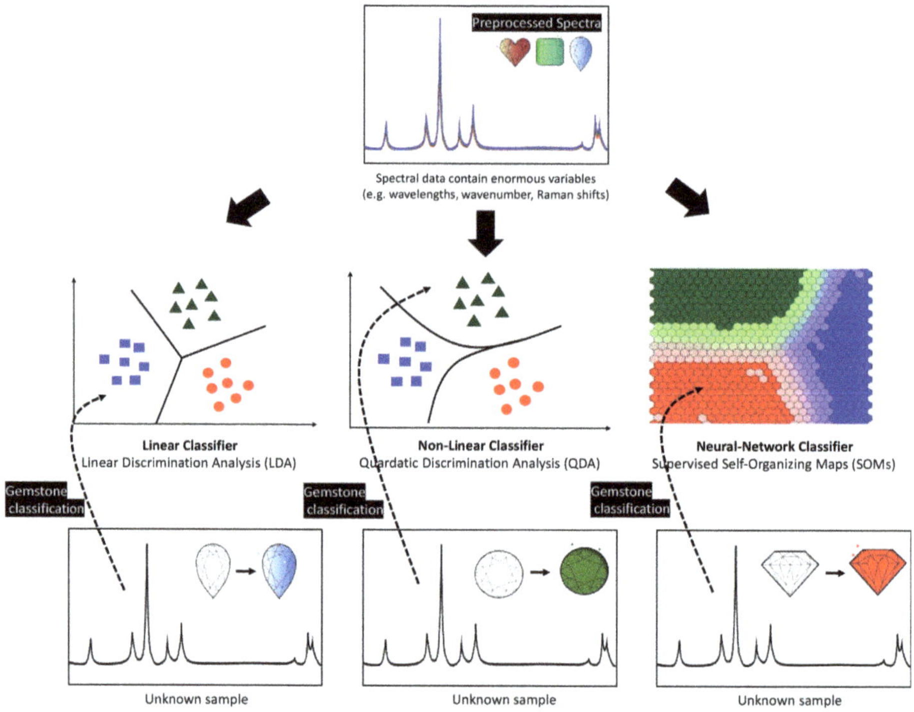

Figure 3.15. Under the supervised pattern recognition approach, the linear, non-linear, and neural network classifier (boundary) of classes of gemstone is generated using LDA, QDA, and supervised SOMs, respectively. To determine a group of an unknown sample, the spectrum of an unknown gemstone is superimposed to the classifier and then applied to the created boundary to determine which group the unknown sample belongs to.

Moreover, feature selection allows the reduction of the dimension of data and leads to more compact models which possibly produce a better generalization ability. The chosen spectral region could be further used to pair with the chemical information of gemstones. This allows us to get more insight of the specific chemical as marker that existed in each type of gemstone. There are two main types of feature selection techniques involving supervised and unsupervised methods. Most of the feature selection methods are developed based on a supervised task, which include the label of pre-defined class in the calculation. On supervised methods, it is divided into filter, wrappers, and embedded methods. Filter-based feature selections use statistical measures to generate the score index representing correlation with the class target. This involves ANOVA, Pearson correlation, and variance thresholding. In the case of the wrapper technique, searching for any subregions provides the best classifier model. It can be forward, backward, and stepwise selection; however, it is computationally expensive. The embedded method combines the advantages of filter (to select subregion based on score index) and wrapper (select subregion which improves model efficiency) [124–126]. Figure 3.16 demonstrates the impact of feature selection on gemstone classification.

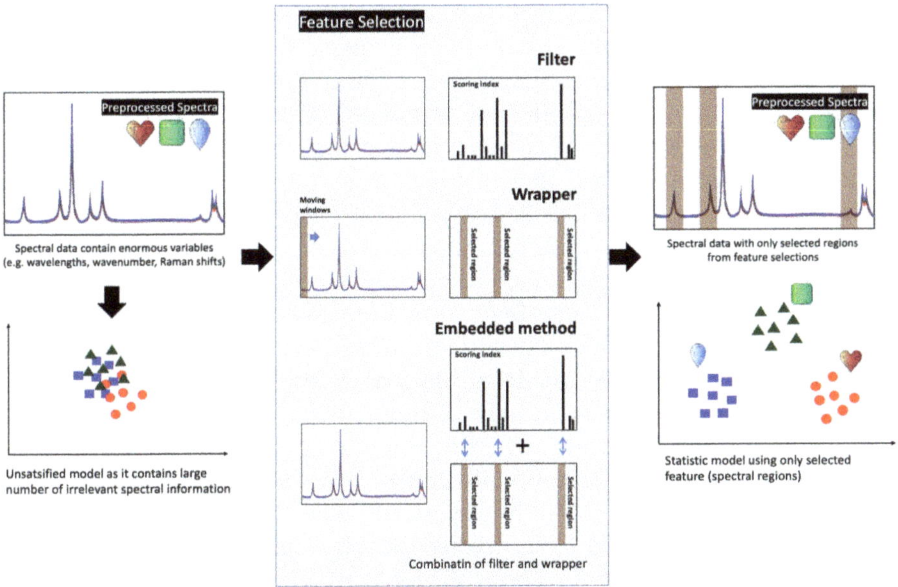

Figure 3.16. Impact of feature selection on the statistical model on gemstone classification.

Applications of machine learning to gemstone classification

In the past decades, researchers have made great efforts to the communities to explore and set up the protocol for gemstone classification and identification using spectroscopic data combined with multivariate data analysis. Pattern recognition will assist in solving the real-world problems from ubiquitous tasks of human decision on the quality and class of gemstones. Prior to data analysis, the good quality of the spectroscopic data should be carefully collected from the novel instrument and the trusted or certificated gemstone samples to be recorded as the references in the database. In the traditional literature, the spectral data information from mineral species which could be relevant in gemology have been published. Fortunately, a large database of Raman spectra acquired from mineral species has constantly been increasing in the last decade. Some databases are free to access via their website (http://www.irug.org/resources/spectral-databases-for-raman) [127–131]. The RRUFF project [127], through the Department of Geosciences, University of Arizona, Tucson, Arizona (http://rruff.info/), offers a complete set of high-quality spectral data including Raman, X-ray diffraction, and electron microprobes from well-characterized minerals. The data provide the standard information for mineralogists, geoscientists, gemmologists, and the general public to identify types of minerals. The project from the Romanian Database of Raman Spectroscopy [128] (Department of Geology, The Alexandru Ioan Cuza University, Iasi, Romania; http://rdrs.uaic.ro/) contains the Raman spectra of individual mineral species including natural gemstone, synthetic gemstone, and artefact samples with several color pigments. Moreover, the software of

Raman Data Search and Storage is also available as the analytical tool for accurate identification of unknown minerals by using the search function to match their sample spectra with the library index data. RASMIN represent a set of Raman spectra obtained from some major mineral groups. The spectra provide the pattern of vibrational peaks of each mineral group. Interestingly, this database delivering the Raman spectra of gemstones in extreme conditions such as high pressure and high temperature had been also carried out. Meanwhile, more extended databases with automatic search engines are also commercially available. In the traditional literatures, the small catalogues of spectral data of gemstone species that are relevant in gemology have been also published elsewhere [132, 133]. Culka *et al* had developed novel instrumentation, namely PSSERS, which is able to effectively suppress the laser-induced fluorescence signal in order to enhance the signal-to-noise ratio. The Raman spectra of 42 minerals (from 59 samples) including chrysoberyl, diamond, emerald, garnet, magnesiotaaffeite-2N′2S (taaffeite), magnesiotaaffeite-6N′3S (musgravite), ruby, and topaz as well as other less precious minerals were collected. The selected gemstone sample represent minerals originated from diverse localities and more color variants.

Due to the fast development in spectrometers with high resolution, both handheld and portable, they provide rapid and fast measurement to acquire the spectra of the samples. Therefore, the usages of the spectral data have received large attention as the standard for the rapid verification, identification, and classification of the precious and semiprecious gemstones. At the starting state, the characteristic peaks on the unknown were manually matched with the reference spectra on the database. Lowry *et al* [134] used the Raman spectra database as a tool to verify whether gem samples were treated with polymer or epoxy. It was observed that different samples (96 semiprecious gemstones) produced the significant different levels of Raman intensity from treated gemstones. Barone *et al* [135] aimed to investigate the potential of micro-Raman spectroscopy in characterizing gems and jewellery from jewellers' collections. The set of red gems in the trading markets were analysed using different laser excitation sources. It was found that the obtained spectra could discriminate rubies from simulant and fakes. Moreover, the useful information of specific spectra regions related to inclusions and treatments was achievably revealed to fulfill a complete characterization of the studies of red gems. Culka *et al* [136] offers the application of the Gem Raman instrument (portable system) to monitor the silicate minerals in the gemstone varieties. It was found that it provided the lesser quality of spectra when compared with the spectra from the standard Raman spectrometer. However, general identification of gemstones was achieved, but it needed spectral modification to classify some gemstone group such as rubies. Jehlicka *et al* [137] used a lightweight portable Raman spectrometer (equally palm-sized device) for fast and unambiguous detection of common gemstones mounted in ancient jewels. This provided the gemstone discrimination based on wavenumbers of the strongest Raman bands of the minerals: quartz and its varieties, beryl varieties (emerald), corundum varieties (sapphire), garnets (almandine, grossular), and diamond as well as aragonite in pearls.

In the past decade, there have been enormous works on the applications of spectrophotometers to classify groups of gemstones because they offer fast and non-destructive analyses of gems *in situ* mounted with jewels. Some jewels are invaluable, which could be possible to bring them to the standard laboratory for practical investigation. Because of this reason, nowadays, a large spectral database of gemstones is available, and most of them are free to access. Therefore, machine learning plays an important role here to match the spectra acquired elsewhere to the spectra on the existing database for gem classification and identification. Culka and Jehlicka [138] examine the possibility of fast and unambiguous identification of various minerals presented in cut form of gemstones using their Raman spectra (acquired by a portable Raman instrument). Figure 3.17 shows a test scenario of a blind study based on the identification of 20 unidentified cut gemstones solely using their Raman spectra combined with the online RRUFF database. The obtained Raman spectra were uploaded into the freeware CrystalSleuth program (machine learning program) to perform searches in the online database of Raman spectra of minerals RRUFF. It was found that 19 out of 20 cut minerals or gemstones were correctly identified based on multiple similar matches of their Raman spectra and the standard Raman spectra in the RRUFF database. The straightforward protocol described in this study suggests that the gemstone discrimination using only a portable Raman spectrometer, and a free online spectral library is highly feasible even for a user not experienced in

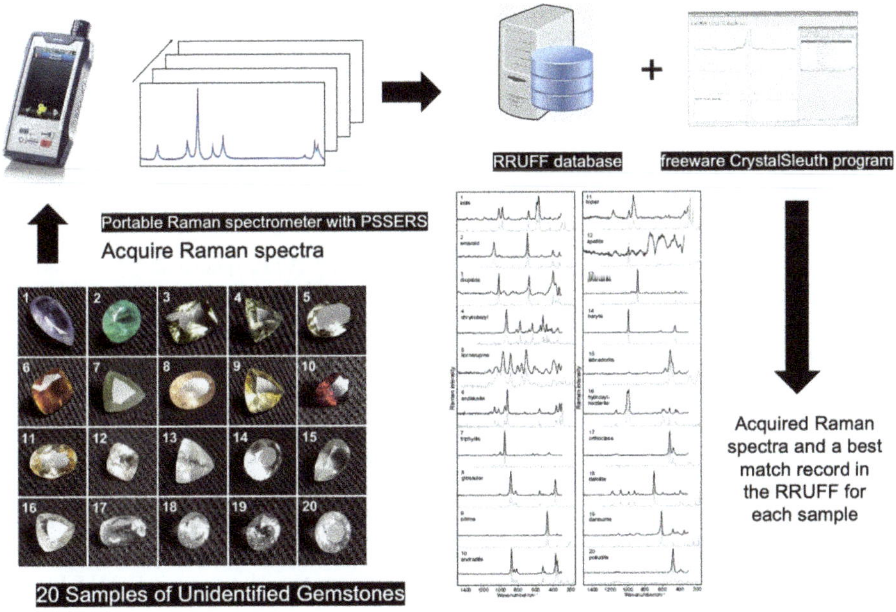

Figure 3.17. Comparison of Raman spectra of 20 unidentified cut gemstones as recorded with a PSSERS and portable Raman spectrometer; all spectra were imported to the database library, and a best match spectrum was recorded in the RRUFF database for each gemstone sample (grey spectra).

Raman spectroscopy and avoids the usual searching for the traditional spectroscopic references in the literatures.

3.10 Conclusions

This chapter presents examples of certain gemstone analyses using the FTIR technique. This FTIR approach can still be used to analyse a wide range of other gemstones. Conclusively, the FTIR technique can be used to classify different types of gemstones as well as distinguish between natural gems, synthetic gems, treated gems, and gem imitations. FTIR spectroscopy, which is an analytical technique that measures a material at a molecular level and produces molecular unique fingerprints, can also aid in tracking the origin of gems. A combination of FTIR spectroscopy and machine learning algorithm sheds the light on numerous applications. The AI application will improve the speed, precision, and accuracy of an analysis technique, which will benefit academic and commercial goals.

References

[1] Chung C, Lee M and Choe E K 2004 Characterization of cotton fabric scouring by FT-IR ATR spectroscopy *Carbohydrate Polym.* **58** 417–20

[2] Gómez-Ordóñez E and Rupérez P 2011 FTIR-ATR spectroscopy as a tool for polysaccharide identification in edible brown and red seaweeds *Food Hydrocoll.* **25** 1514–20

[3] Belfer S, Fainchtain R, Purinson Y and Kedem O 2000 Surface characterization by FTIR-ATR spectroscopy of polyethersulfone membranes-unmodified, modified and protein fouled *J. Membr. Sci.* **172** 113–24

[4] Kiwi J and Nadtochenko V 2005 Evidence for the mechanism of photocatalytic degradation of the bacterial wall membrane at the TiO_2 interface by ATR-FTIR and laser kinetic spectroscopy *Langmuir* **21** 4631–41

[5] Baruch M F, Pander J E, White J L and Bocarsly A B 2015 Mechanistic insights into the reduction of CO_2 on tin electrodes using *in situ* ATR-IR spectroscopy *ACS Catal.* **5** 3148–56

[6] Fieldson G T and Barbari T A 1993 The use of FTIR-ATR spectroscopy to characterize penetrant diffusion in polymers *Polymer* **34** 1146–53

[7] Fujiwara M, Chow P S, Ma D L and Braatz R D 2002 Paracetamol crystallization using laser backscattering and ATR-FTIR spectroscopy: metastability, agglomeration, and control *Cryst. Growth Des.* **2** 363–70

[8] Baranska M, Schütze W and Schulz H 2006 Determination of lycopene and β-carotene content in tomato fruits and related products: comparison of FT-Raman, ATR-IR, and NIR spectroscopy *Anal. Chem.* **78** 8456–61

[9] Anjos O, Campos M G, Ruiz P C and Antunes P 2015 Application of FTIR-ATR spectroscopy to the quantification of sugar in honey *Food Chem.* **169** 218–23

[10] Napavichayanun S, Amornsudthiwat P, Pienpinijtham P and Aramwit P 2015 Interaction and effectiveness of antimicrobials along with healing-promoting agents in a novel biocellulose wound dressing *Mater. Sci. Eng.* C **55** 95–104

[11] Simon M, Vianello A, Shashoua Y and Vollertsen J 2021 Accelerated weathering affects the chemical and physical properties of marine antifouling paint microplastics and their identification by ATR-FTIR spectroscopy *Chemosphere* **274** 129749

[12] Peets P, Leito I, Pelt J and Vahur S 2017 Identification and classification of textile fibres using ATR-FT-IR spectroscopy with chemometric methods *Spectrochim. Acta* A **173** 175–81
[13] Pienpinijtham P, Thammacharoen C, Naranitad S and Ekgasit S 2018 Analysis of cosmetic residues on a single human hair by ATR FT-IR microspectroscopy *Spectrochim. Acta* A **197** 230–6
[14] Thongnopkun P 2005 Characterization of some varieties of gemstone by FT-IR spectroscopy *PhD Thesis* Chulalongkorn University, Bangkok
[15] Thongnopkun P and Ekgasit S 2005 FTIR spectra of faceted diamonds and diamond simulants *Diam. Relat. Mater.* **14** 1592–9
[16] Tretyakova L I, Reshetnyak N B and Tretyakova V Y 1997 Combined spectroscopic method for non-destructive gem identification *J. Gemmol.* **25** 532–9
[17] Ferrer N and Jogués-Carulla J M 1996 Characterization study of cut gem diamond by IR spectroscopy *Diam. Relat. Mater.* **5** 598–602
[18] Thongnopkun P and Ekgasit S 2006 Attenuated total reflection Fourier transform infrared spectra of faceted diamonds *Anal. Chim. Acta* **576** 130–5
[19] Guiliano M, Asia L, Onoratini G and Mille G 2007 Applications of diamond crystal ATR FTIR spectroscopy to the characterization of ambers *Spectrochim. Acta* A **67** 1407–11
[20] Ekgasit S and Thongnopkun P 2005 Transflectance spectra of faceted diamonds acquired by infrared microscopy *Appl. Spectrosc.* **59** 1160–5
[21] Breeding C M and Shigley J E 2009 The 'type' classification system of diamonds and its importance in gemology *Gems Gemol.* **45** 96–111
[22] Fritsch E and Phelps A W 1993 Type IIb diamond thin films deposited onto near-colorless natural gem diamonds *Diam. Relat. Mater.* **2** 70–4
[23] Fritsch E, Hainschwang T, Massi L and Rondeau B 2007 Hydrogen-related optical centers in natural diamond: an update *New Diamond Front. Carbon Technol.* **17** 63–89
[24] Hainschwang T, Notari F and Pamies G 2020 The origin of color of 1330 nm center diamonds *Diam. Relat. Mater.* **110** 108151
[25] Iakoubovskii K and Adiaenssens G J 2002 Optical characterization of natural argyle diamond *Diam. Relat. Mater.* **11** 125–31
[26] Lu H C and Cheng B M 2011 Analysis of nitrogen defects in diamond with VUV photoluminescence *Anal. Chem.* **83** 6539–44
[27] Overton T W and Shigley J E 2008 A history of diamond treatment *Gems Gemol.* **44** 32–55
[28] Petkewich A 2003 Finding a real gem *Anal. Chem.* **75** 71A–4A
[29] Jayaraman A 2000 A Brief overview of gem materials: natural and synthetic *Curr. Sci.* **79** 1555–65
[30] Read P G 1995 *Gemmology* (Oxford: Butterworth-Heinemann)
[31] Shigley J E 2000 Treated and synthetic gem materials *Curr. Sci.* **79** 1566–71
[32] Eaton-Magana S, Shigley J E and Breeding C M 2017 Observations on HPHT-grown synthetic diamond: a review *Gems Gemol.* **53** 262–84
[33] Peretti A, Herzog F, Bieri W, Alessandri M, Günther D, Frick D A, Cleveland E, Zaitsev A M and Deljanin B 2014 New generation of synthetic diamonds reaches the market, Part A: CVD-grown blue diamonds *Contrib. Gemol.* **14** 1–16
[34] McNamara K M, Williams B E, Gleason K K and Scruggs B E 1994 Identification of defects and impurities in chemical-vapor-deposited diamond through infrared spectroscopy *J. Appl. Phys.* **76** 2466–72

[35] Martynova T V, Polushin N I, Laptev A I and Maslov A L 2020 Express-analysis of nitrogen content in CVD-diamonds by FTIR-spectrometry *IOP Conf. Ser.: Mater. Sci. Eng.* **919** 022049

[36] Wang W, D'Haenens-Johansson U F S, Johnson P, Moe K S, Emerson E, Newton M E and Moses T M 2012 CVD synthetic diamonds from gemesis corp *Gems Gemol.* **48** 80–97

[37] Martineau P M, Lawson S C, Taylor A J, Quinn S J, Evans D J F and Crowder M J 2004 Identification of synthetic diamond grown using chemical vapor deposition *Gems Gemol.* **40** 2–5

[38] Deljanin B, Herzog F, Bieri W, Alessandri M, Gunther D, Frick D A, Cleveland E, Zaitsev A M and Peretti A 2014 New generation of synthetic diamonds reaches the market part B: identification of treated CVD-grown pink diamonds from Orion (PDC) *Contrib. Gemol.* **14** 21–40

[39] Zaitsev A M, Deljanin B, Peretti A, Alessandri M and Bieri W 2014 New generation of synthetic diamonds reaches the market. Part C: origin of yellow color in CVD-grown diamonds and treatment experiments *Contrib. Gemol.* **14** 41–55

[40] Arantes T M, Santos N M, Azevedo A F, Baldan M R and Ferreira N G 2014 Fourier transform infrared spectroscopic study of boron-doped micro/nano/ultra nanocrystalline diamond prepared by chemical vapor deposition *Mater. Sci. Forum* **802** 140–5

[41] Meng Y, Yan C, Lai J, Krasnicki S, Shu H, Yu T, Liang Q, Mao H and Hemley R J 2008 Enhanced optical properties of chemical vapor deposited single crystal diamond by low-pressure/high-temperature annealing *Proc. Natl Acad. Sci.* **105** 17620–5

[42] Collins A T 2003 The detection of colour-enhanced and synthetic gem diamonds by optical spectroscopy *Diam. Relat. Mater.* **12** 1976–83

[43] Fisher D and Spits R 2000 A spectroscopic evidence of GE POL HPHT-treated natural type IIa diamonds *Gems Gemol.* **36** 42–9

[44] Weerdt F D and Collins A T 2008 Determination of the C defect concentration in HPHT annealed type IaA diamonds from UV–VIS absorption spectra *Diam. Relat. Mater.* **17** 171–3

[45] Wang M, Ghi S, Yuan J C C, Han W and Bai Q 2018 Spectroscopic characteristics of treated-color natural diamonds *J. Spectroscop.* **2018** 8153941

[46] Pauling L and Hendricks S B 1925 The crystal structure of hematite and corundum *J. Am. Chem. Soc.* **47** 781–90

[47] Hughes R W 1997 *Ruby & Sapphire* (Boulder, CO: RWH Publishing)

[48] Hughes R W 2014 *Ruby & Sapphire. A Collector's Guide* (Bangkok: Gems and Jewelry Institute of Thailand)

[49] Belt R F 1967 Hydrothermal ruby: infrared spectra and X-ray topography *J. Appl. Phys.* **38** 2688–9

[50] Beran A Trace hydrogen in Verneuil-grown corundum and its color varieties on IR spectroscopic study *Eur. J. Mineral.* **3** 971–5

[51] Smith C P 1995 A contribution to understanding the infrared spectra of rubies from Mong Hsu, Myanmar *J. Gemmol.* **24** 321–35

[52] Gübelin E J 1982 New synthetic rubies made by Professor P.O. Knischka *Gems Gemol.* **18** 165–8

[53] Kammerling R C, Koivula J I and Fritsch E 1994 An examination of Chatham flux-grown synthetic pink sapphires *J. Gemmol.* **24** 149–54

[54] Kane R E 1982 The gemological properties of Chatham fluxgrown synthetic orange sapphire and synthetic blue sapphire *Gems Gemol.* **18** 140–53

[55] Kuznetsov V A and Shternberg A A 1967 Crystallization of ruby under hydrothermal conditions *Sov. Phys. Crystallogr.* **12** 280–5

[56] Laudise R A and Ballman A A 1958 Hydrothermal synthesis of sapphire *J. Am. Chem. Soc.* **80** 2655–7

[57] Thomas V G, Mashkovtsev R I, Smirnov S Z and Maltsev V S 1997 Tairus hydrothermal synthetic sapphires doped with nickel and chromium *Gems Gemol.* **33** 188–202

[58] Peretti A and Smith C P 1993 A new type of synthetic ruby on the market, offered as hydrothermal rubies from Novosibirsk *Aust. Gemologist* **18** 149–56

[59] Volynets F K, Sidorova Y A and Stsepuro N A 1974 OH-groups in corundum crystals grown by the Verneuil technique *J. Appl. Spectrosc.* **17** 1626–8

[60] Muller R and Gunthard H H 1966 Spectroscopic study of the reduction of nickel and cobalt ions in sapphire *J. Chem. Phys.* **44** 365–73

[61] Eigenmann K and Günthard H H 1971 Hydrogen incorporation in doped alpha-Al_2O_3 by high temperature redox reaction *Chem. Phys. Lett.* **12** 12–5

[62] Moon A R and Phillips M R 1991 Detec clustering in H, Ti: α-Al_2O_3 *J. Phys. Chem. Solids* **52** 1087–99

[63] Moon A R and Phillips M R 1994 Detect clustering and color in Fe, Ti:α-Al_2O_3 *J. Am. Ceram. Soc.* **77** 356–67

[64] Phlayrahan A, Monarumit N, Lhuaamporn L, Satitkune S and Wathanakul P 2019 Spectroscopic investigation of properties of blue sapphire samples depending on heating conditions *J. Appl. Spectrosc.* **86** 810–6

[65] Phlayrahan A, Monarumit N, Satitkune S and Wathanakul P 2018 Role of Ti content on the occurrence of the 3309-cm^{-1} peak in FTIR absorption spectra of ruby samples *J. Appl. Spectrosc.* **85** 385–90

[66] Jaliya R G C, Dharmaratne P G R and Wijieskara K B 2020 Characterization of heat treated geuda gemstones for different furnance condition using FTIR, XRD and UV–Visible spectroscopy methods *Solid Earth Sci.* **5** 282–9

[67] Sun T T and Ying L H 2014 Cobalt-doped glass-filled treated blue sapphire *IGCL Newsletter* http://icglabs.org/wp-content/uploads/2014/12/ICGLNewsletter-2014FA.pdf

[68] Leelawatanasuk T, Atichat W, Pisutha-Arnond V, Wattanakul P, Ounorn P and Manorotkul W *et al* 2013 Cobalt-doped glass-filled sapphires: an update *Aust. Gemmologist* **25** 14–20

[69] Leelawatanasuk T, Susawee N, Promwongnam S and Atsawatanapirom N 2015 Green lead-glass-filled sapphire *J. Gemmol.* **34** 420–7

[70] Lalous G 2013 Cobalt-doped composite sapphires: a blessing or a curse? *Gems Gemol.* **49** 176

[71] Huong L T, Hofmeister W, Häger T, Karampelas S and Kien N D 2014 A preliminary study on the separation of natural and synthetic emeralds using vibrational spectroscopy *Gems Gemol.* **50** 287–92

[72] Reshma B, Sakthivel R and Mohanty J K 2016 Characterization of low grade natural emerald gemstone *J. Geol. Geophys.* **6** 1000271

[73] Schmetzer K, Kiefert L, Bernhardt H and Beili Z 1997 Characterization of Chinese hydrotermal synthetic emerald *Gems Gemol.* **33** 276–91

[74] Bellatreccia F, Ventura G D, Piccinini M and Grubessi O 2008 Single-crystal polarised-light FTIR study of an historical synthetic water-poor emerald *Neues Jahrbuch für Mineralogie - Abhandlungen* **185** 11–6

[75] Mashkovtsev R I, Thomas V G, Fursenko D A, Zhukova E S, Uskov V V and Gorshunov B P 2016 FTIR spectroscopy of D_2O and HDO molecules in the c-axis channels of synthetic beryl *Am. Mineral.* **101** 175–80
[76] Kiefert L, Hanni H A, Chalain J P and Weber W 1999 Identification of filler substances in emeralds by infrared and Raman spectroscopy *J. Gemmol.* **26** 501–20
[77] Qiao X, Zhou Z, Schwarz D T, Qi L, Gao J, Nong P, Lai M, Guo K and Li Y 2019 Study of the differences in infrared spectra of emerald from different mining areas and the controlling factors *Can. Mineral.* **57** 65–79
[78] Ventura G D, Radica F, Bellatreccia F, Freda C and Guidi M C 2015 Speciation and diffusion profiles of H_2O in water-poor beryl: comparison with cordierite *Phys. Chem. Miner.* **42** 735–45
[79] Ventura G D, Radica F, Bellatreccia F, Cavallo A, Cinque G, Tortora L and Behrens H 2015 FTIR imaging in diffusion studies: CO_2 and H_2O in a synthetic sector-zoned beryl *Front. Earth Sci.* **3** 33
[80] Wood D L and Nassau K 1986 The characterization of beryl and emerald by visible and infrared absorption spectroscopy *Am. Mineral.* **53** 777–800
[81] Gervais F, Piriou B and Cabannes F 1972 Anharmonicity of infrared vibration modes in beryl *Physica Status Solidi* B **51** 701–12
[82] Aitsicchio C, Grtibessi O and Zecchin P 1994 Infrared spectroscopy and crystal chemistry of the beryl group *Can. Mineral.* **32** 55–68
[83] Charoy B, Donato P, Barres O and Pinto-Coelho C 1996 Channel occupancy in an alkali-poor beryl from Serra Branca (Goias, Brazil): spectroscopic characterization *Am. Mineral.* **81** 395–403
[84] Branca C, Arcovito A, Cosio E, Interdonato M, Sabatino G, Wanderlingh U and D'Angelo G 2019 Combining Fourier transform infrared and Raman spectroscopies with Gaussian deconvolution: an improved approach for the characterization of emeralds *J. Raman Spectrosc.* **51** 693–701
[85] Aurisicchio C, Corami A, Ehrman S, Graziani G and Cesaro S N 2006 The emerald and gold necklace from oplontis, Vesuvian Area, Naples, Italy *J. Archaeolog. Sci.* **33** 725–34
[86] Karampelas S, Al-Shaybani B, Mohamed F, Sangsawong S and Al-Alawi A 2019 Emeralds from the most important occurrences: chemical and spectroscopic data *Minerals* **9** 561
[87] Johnson M L 2007 Durability testing of filled emeralds *Gems Gemol.* **43** 120–37
[88] Nassau K 1978 The origins of color in minerals *Am. Mineral.* **63** 219–29
[89] Fritsch E and Rossman G R 1988 An update on color in gems. Part 3: color caused by band gaps and physical phenomena *Gems Gemol.* **24** 81–102
[90] Crowningshield R, Hurlbzzt C and Fryer C W 1986 A simple procedure to separate natural from synthetic amethyst on the basis of twinning *Gems Gemol.* **22** 130–93
[91] Karampelas S, Fritsch E, Zorba T and Paraskevopoulos K M 2011 Infrared spectroscopy of natural vs. synthetic amethyst: an update *Gems Gemol.* **47** 196–201
[92] Karampelas S, Fritsch E, Zorba T, Paraskevopoulos K M and Sklavounos S 2005 Distinguishing natural from synthetic amethyst: the presence and shape of the 3595 cm^{-1} peak *Mineral. Petrol.* **85** 45–52
[93] Staats P A and Kopp O C 1974 Studies on the origin of the 3400 cm^{-1} region infrared bands of synthetic and natural α-quartz *J. Phys. Chem. Solids* **35** 1029–33

[94] Balitsky V S, Balitsky D, Bondarenko G V and Balitskaya O V 2004 The 3543 cm^{-1} Infrared absorption band in natural and synthetic amethyst and its value in identification *Gems Gemol.* **40** 146–61
[95] Saikia B J, Parthasarathy G and Sarmah N C 2008 Fourier transform infrared spectroscopic estimation of crystallinity in SiO_2 based rocks *Bull. Mater. Sci.* **31** 75–9
[96] Breeding C M 2009 Using LA-ICP-MS analysis for the separation of natural and synthetic amethyst and citrine News from Research, July 31, 2009 http://www.gia.edu/research-resources/news-from-research
[97] Pinto L C B M, Righi A, Lameiras F S, Araujo F G S and Krambrock K 2011 Origin of the color in cobalt-doped quartz *Phys. Chem. Miner.* **38** 623–9
[98] Nunes E H M and Lameiras F S 2005 The optical absorption of gamma irradiated and heat treated natural quarts *Mater. Res.* **8** 305–08
[99] Nunes E H M, Lameiras F S, Houmard M and Vasconcelos W L 2013 Spectroscopic study of natural quartz samples *Radiat. Phys. Chem.* **90** 79–86
[100] Tan T L, LL N G and Lim L C 2013 Studies on nephrite and jadeite jades by FTIR spectroscopic technique and Raman technique *COSMOS* **9** 47–56
[101] Maia Y, Wanga R and Cui H 2021 Study on the blackening mechanism of buried cinnabar within ancient Chinese jades *J. Cult. Heritage* **49** 164–73
[102] Yang C M O and Hansheng L Review of recent studies on black jadiete jade *J. Gemmol.* **26** 417–24
[103] Manrique-Ortega M D, Mitrani A, Casanova-González E, Jiménez-Galindo L A and Ruvalcaba-Sil J L 2019 Methodology for the non-destructive characterization of jadeite-jade for archaeological studies *Spectrochim. Acta* A **217** 294–309
[104] Korybska-Sadło I, Gil G, Gunia P, Horszowski M and Sitarz M 2018 Raman and FTIR spectra of nephrites from the Złoty Stok and Jordanow Slaski (the sudetes and fore-sudetic block, SW Poland) *J. Mol. Struct.* **1166** 40–7
[105] Zhang J, Lu T and Chen H 2013 Characteristics of coated jadeite jade *Gems Gemol.* **49** 246–51
[106] Lai L T 2014 Polymer-impregnated aventurine quartz, a new imitation of ice jade *Gems Gemol.* **50** 90–1
[107] Liu Y, Lu T, Wang M, Chen H, Shen M, Ke J and Zhang B 2009 Identification of dyed jadeite using visible reflection spectra *J. Gemmol.* **31** 181–4
[108] Wang X and Guo Y 2021 The impact of trace metal cations and absorbed water on colour transition of turquoise *R. Soc. Open Sci.* **8** 201110
[109] Cejka J, Sejkora J, Macek I, Malikova R, Wang L, Scholz R, Xi Y and Frost R L 2015 Raman and infrared spectroscopic study of turquoise minerals *Spectrochim. Acta* A **149** 173–82
[110] Abdu Y A, Hull S A, FayeK M and Hawthorne F C 2011 The turquoise-chalcosiderite Cu $(Al,Fe^{3+})_6(PO_4)_4(OH)_8{\cdot}4H_2O$ solid-solution series: a Mössbauer spectroscopy, XRD, EMPA, and FTIR study *Am. Mineral.* **96** 1433–42
[111] Moe K S, Moses T M and Johnson P 2007 Polymer-impregnated turquoise *Gems Gemol.* **43** 149–51
[112] Fritsch E, McClure S F, Ostrooumov M, Andres Y, Moses T, Koivula J I and Kammerling R C 1999 The identification of Zachery-treated turquoise *Gems Gemol.* **35** 4–16
[113] Kadikova I F, Morozova E A, Yuryeva T V, Grigorieva I A and Yuryev V A 2020 Glass depolymerization in the process of long-term corrosion: a study of deteriorating semi-opaque turquoise glass beads using micro-FTIR spectroscopy *Mater. Res. Express* **7** 025203

[114] Brereton R G 2003 *Chemometrics: Data Analysis for the Laboratory and Chemical Plant* (New York: Wiley)

[115] Brereton R G 2009 *Chemometrics for Pattern Recognition* (New York: Wiley)

[116] Brereton R G 2007 *Applied Chemometrics for Scientists* (New York: Wiley)

[117] Araujo P W and Brereton R G 1996 Experimental design. I: screening *TrAC Trends Anal. Chem.* **15** 26–31

[118] Araujo P W and Brereton R G 1996 Experimental design. II: optimization *TrAC Trends Anal. Chem.* **15** 63–70

[119] Lloyd G R, Wongravee K, Silwood C L, Grootveld M and Brereton R G 2009 Self-organising maps for variable selection: application to human saliva analysed by nuclear magnetic resonance spectroscopy to investigate the effect of an oral healthcare product *Chemometr. Intell. Lab. Syst.* **98** 149–61

[120] Wongravee W, Lloyd G R, Silwood C J, Grootveld M and Brereton R G 2010 Supervised self organising maps for classification and determination of potentially discriminatory variables: illustrated by application to nuclear magnetic resonance metabolomic profiling *Anal. Chem.* **82** 628–38

[121] Lloyd G R, Brereton R G and Duncan J C 2008 Self organising maps for distinguishing polymer groups using thermal response curves obtained by dynamic mechanical analysis *Analyst* **133** 1046–59

[122] Lloyd G R, Brereton R G, Faria R and Duncan J C 2007 Learning vector quantization for multiclass classification: application to characterization of plastics *Analyst* **47** 1553–63

[123] Brereton R G and Lloyd G 2010 Support vector machines for classification and regression *Analyst* **135** 230–67

[124] Zhao Z A and Liu H 2012 *Spectral Feature Selection for Data Mining* (London: Chapman and Hall/CRC)

[125] Guyon I 2003 An introduction to variable and feature selection *J. Mach. Learn. Res.* **3** 1157–82

[126] Saeys Y, Inza I and Larranaga P A 2007 Review of feature selection techniques in bioinformatics *Bioinformatics* **23** 2507–17

[127] Lafuente B, Downs R T, Yang H and Stone N 2015 The power of databases: the RRUFF project *Highlights of Mineralogical Crystallography* ed T Armbruster and R M Danisi (Berlin: De Gruyter) pp 1–30

[128] Buzgar N, Apopei A I and Buzatu A 2021 Romanian Database of Raman Spectroscopy http://rdrs.uaic.ro/

[129] Handbook of Minerals Raman Spectra 2021 http://ens-lyon.fr/LST/Raman/ index.php

[130] RASMIN 2021 Raman Spectra Database of Minerals and Inorganic Materials http://riodb.ibase.aist. go.jp/rasmin/

[131] Raman spectra database 2021 http://dst.unisi.it/geofluids/raman/spectrum_frame.htm

[132] Culka A and Jehlicka J A A 2019 Database of Raman spectra of precious gemstones and minerals used as cut gems obtained using portable sequentially shifted excitation Raman spectrometer *J. Raman Spectrosc.* **50** 262–80

[133] Bersani D, Azzi G, Lambruschi E, Barone G, Mazzoleni P, Raneri P, Raneri S, Longobardo U and Lottici P P 2014 Characterization of emeralds by micro-Raman spectroscopy *J. Raman Spectrosc.* **45** 1293–300

[134] Lowry S, Wiebolt D, Dalrymple D, Jasinevicius R and Downs R T 2009 The use of a Raman spectral database of minerals for the rapid verification of semiprecious gemstones *Spectroscopy (Santa Monica)* **24** 1–7

[135] Barone G, Bersani D, Lottici P P, Mazzoleni P, Raneri S and Longobardo U 2016 Red gemstone characterization by micro-Raman spectroscopy:the case of rubies and their imitations *J. Raman Spectrosc.* **47** 1534–9
[136] Culka A, Jaroslav H and Jehlicka J A 2016 Gem and mineral identification using GL gem Raman and comparison with other portable instruments *Appl. Phys.* A **122** 959
[137] Jehlička J, Culka A, Baštov M, Bašta P and Kuntoš J 2016 The ring monstrance from the Loreto treasury in Prague: handheld Raman spectrometer for identification of gemstones *Phil. Trans. R. Soc.* A **374** 20160042
[138] Culka A and Jehlička J 2019 Identification of gemstones using portable sequentially shifted excitation Raman spectrometer and RRUFF online database: a proof of concept study *Eur. Phys. J. Plus* **134** 130

IOP Publishing

Artificial Intelligence and Spectroscopic Techniques for Gemology Applications

Ashutosh Kumar Shukla

Chapter 4

A ruby stone grading inspection using an optical tomography system

Syarfa Najihah Raisin, Juliza Jamaludin and Fatinah Mohd Rahalim

This paper proposes the ruby stone grading inspection using an optical tomography system approach. The Linear Back Projection (LBP) algorithm was applied to reconstruct the three-dimensional images of ruby stones with refractive index values of 1.762 and 1.770. From these image reconstructions, the transparency of ruby stones can be analyzed through ANOVA statistical testing, hence constituting an evaluation of the stones' grading. Results from the statistical analysis successfully demonstrate the capability of the optical tomography method with the laser as a transmitter in distinguishing between two difference refractive index values of ruby stones.

4.1 Introduction

For centuries, the world's supply of rubies has been dominated by two sources: modern-day Sri Lanka and the Mogok region in Myanmar. From the late 1800s to the 1990s, the border areas of what is now Thailand and Cambodia also produced high-quality gems, although this deposit only became dominant in the 1960s [1]. High demand in the global market has led to the creation and proliferation of counterfeited or imitation ruby stones. The grading valuation of ruby stones is categorized by its color, transparency, and presence of blemishes.

In recent technological advancements, there are a number of gemology tools that are used to examine the optical properties of synthetic and non-synthetic ruby stones, tools such as the loupe, microscope, and dichroscope [2, 3]. Nevertheless, these tools are highly dependent on human visual assessments made by highly experienced professionals. However, these methods are not standardized and may lead to inaccurate grading valuations due to their dependency on pictorial inspection [2, 3]. According to Jiaqi [3], the chemical and physical characteristics and inner

doi:10.1088/978-0-7503-3927-8ch4

structures of gemstones are typically analyzed using microscopic observations, polarizing microscopy, and Raman spectroscopy. When a cambered gemstone is illuminated by a point light, the light reflects off of its surface and forms bright light bands on top of the convex gemstone. Unfortunately, the aforementioned technique is not accurate. Moreover, it is not a consistent method because it is inherently dependent on the gemstones' inner structure.

Another preceding study conducted by Xin Pan [4] employed statistics for the purpose of gemstone grading valuation. The said study was based upon the statistical analysis of the different colors of 120 red jadeite jade samples. The samples were tested using an X-Rite SP62 integral spherical spectrophotometer. The K-means clustering analysis method in the IBM SPSS 22.0 statistical analysis software was then used to grade the color of the red jadeite jade samples. The Fisher discriminant analysis method was additionally utilized to verify the feasibility of the classifications and to also find an objective and effective color grade evaluation method. Finally, the red jadeite jade samples were categorized into five grades: Fancy Vivid, Fancy Deep, Fancy Intense, Fancy Dark, and Fancy [4]. This technique has its own set of limitations because the result is based on color and not a specific quantitative value of the gemstone itself. Here, a quantitative grading value signifies a classification of gemstone based on the light intensity values that it reflects, which are then received by a sensor. Therefore, the current state of knowledge in this topic suggests a gap exists wherein no systematic and standardized technique exists that can quantitatively measure the grading valuation of ruby stones. Based on the aforementioned points, the proposed research aims to introduce the ruby stone grading inspection using an optical tomography system approach.

This research intends to apply the optical tomography technique to investigate the transparency of ruby stone. This approach is non-invasive, non-intrusive, and has the capability to accurately determine the optical properties of objects [5]. The charge-coupled device (CCD) sensor was selected as the optoelectronic sensor to integrate with the optical tomography technique in inspecting the ruby stone grading transparency. A number of preceding research on optical tomography systems that made use of CCDs were conducted by Jamaludin [6–10]. The outcomes of these studies showed that the Sony ILX551A CCD works effectively with a low power laser diode, which is a red-light source. This CCD optical tomography system was found to be capable in identifying the existence of objects with different transparencies and blemishes in crystal-clear water. The classification of object transparency and blemishes was done by analyzing the light intensity received by the CCD optical tomography system after light has penetrated the object. The basic principle used by the abovementioned research is measuring the light intensity received by the CCD [7].

This proposed study on the ruby stone grading inspection using an optical tomography system approach will develop a new method to determine the quantitative grading values of ruby stones for potential use by gemologists and players in the precious stones and manufacturing industries (drilling and cutting) as well as in medical and dentistry applications. Light reflected from the ruby stone strikes the CCD surface. This light source is then converted into an electrical signal, which is in the form of voltage output values [10]. The CCD voltage output represents the light intensity value of the ruby stone. The CCD pixels then gather the photon charges emitted by the light source. The aforementioned constitutes the hypothesis of this research; specifically, the more photon charges that are received by the CCD, the higher the transparency of the studied ruby stones [11].

4.2 Methodology

The CCD sensor has a very special design architecture that involves more than a thousand very small pixels that are photosensitive. The pixels of a CCD can measure just micrometers in size. This enables many thousands of such pixels to fit on just one device. This type of sensor offers benefits such as low noise interference, fast detection speed, and high resolution [9, 10, 12]. As previously mentioned, light that strikes the CCD surfaces will be converted into electrical signals in the form of voltage output values. The voltage output from the CCD represents the actual light intensity level. The CCD's pixels collect photon charges that are received from the light source. The more photon charges received by the CCD, the higher the clarity of the gemstone being observed [9].

A prior study utilized a linear CCD sensor and a low power laser diode to capture an image of a static solid rod, glass rod, and transparent hollow straw in crystal-clear water. The tomographic results show that the CCD is able to discern the different opacity levels of static rods in crystal-clear water [7, 8, 13]. The current study shall apply the aforementioned procedures to determine the quantitative grading value of ruby stones based on the light intensity ratio and CCD voltage output values. The data obtained will then be used to produce a mathematical expression that shows the relationship between the light intensity received by the CCD after it has penetrated the ruby stone with its clarity and, consequently, its quantitative grading valuation.

Another study [8] focused on the application of CCD sensors in an optical tomography system to detect matter in crystal-clear water. Based on the findings obtained, the authors indicated that this method successfully enhanced the CCD's ability to detect objects with low opacities. In that study, the images were reconstructed based on an orthogonal projection of lasers and CCDs within a pipeline, and this research applies the ruby stone, CCD, and laser arrangement as illustrated in figure 4.1 [14].

The LabVIEW software was used for three-dimensional image reconstruction by implementing the filtered Linear Back Projection (LBP) method. The analysis involves different types of objects including the solid rod, transparent rod, and air bubbles [15]. The higher the opacity of an object, then the higher the color intensity introduced. In addition, this study also validates the difference in the 3D images with and without the filtering LBP approach [15]. The appearance and locations of static objects were visualized and compared.

4.2.1 Mathematical expression

Figure 4.2 shows the illustration of light striking through ruby stone and CCD. It shows the light passing through two different mediums, namely the air and the ruby

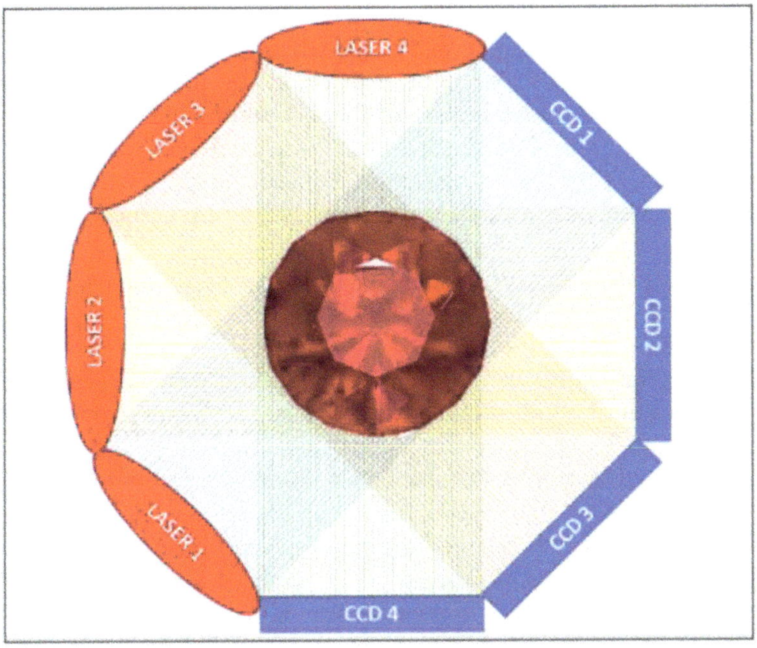

Figure 4.1. Ruby stone, CCD, and laser light source arrangement.

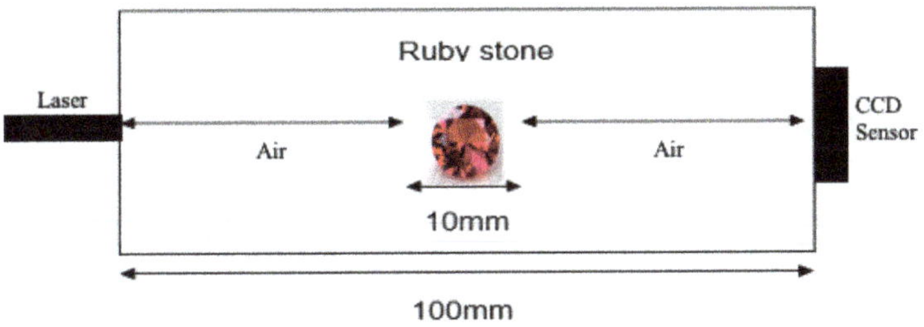

Figure 4.2. Illustration of light striking through ruby stone and CCD.

stone. As light passes through a transparent and translucent object, light attenuation will occur due to light absorption, light reflectance, and light scattering.

Due to the overly complex mathematical model, light scattering and the diffraction effect were elected to be disregarded in this study's calculations. In fact, the wavelength of the incident light used (laser light) is too small compared to the diameter of the object of interest [16]. Based on this, it can be surmised that there are two types of algorithms that are involved in this simulation, specifically light attenuation due to light absorption and light reflectance. Light attenuation occurs when light passes through one medium to another medium, whereby the attenuation phenomena is caused by absorption. The three mediums possess different values for the coefficient α, which will be used in the Beer–Lambert Law as shown in [15].

$$I_{\text{out}} = I_{\text{in}}e^{-\alpha x} \tag{4.1}$$

In the above formula, the output of the light intensity is multiplied by the exponential attenuation of medium density when light passes through it and in which x is the distance of the light transverse. In this study, the x value is manipulated, whereas α is a constant that is represented by the attenuation coefficient of ruby stones. The linear attenuation coefficient of the ruby stone is approximated to that of the linear attenuation coefficient of glass, which is 0.003 [17, 18]. This research will investigate two different types of image reconstruction systems, namely the following:

1. Image reconstruction for System A (Laser–Ruby Stone–CCD)
2. Image reconstruction for System B (Ruby Stone–CCD)

For System A, light from the red laser will strike onto the air–ruby stone–CCD, which will then entail light attenuation because of absorption and reflection. Meanwhile, System B depends on the light reflections from the ruby stone itself that are then detected by the CCD, which only involves light attenuation due to light reflection. The energy of photons in light will decrease when light goes through different mediums or when it passes through adjacent objects. The light reflectance equation is shown in equation (4.2).

$$I_{\text{Final reflection1}} = I_{\text{initial}} - \left[I_{\text{initial}} \left(\frac{n_2 - n_1}{n_2 + n_1} \right)^2 \right] \tag{4.2}$$

From equation (4.2), n_2 refers to the transmitted refractive index, and n_1 is the incidence refractive index [17]. This research involves the selection and analysis of ruby stones with refractive indices (RI) of $n_{\text{ruby stone}} = 1.762$ and 1.770, respectively [19]. By using figure 4.5 as a reference, the transmitted refractive index is represented by the $n_{\text{ruby stone}}$; meanwhile, the incidence refractive index is depicted by $n_{\text{air}} = 1$. The mathematical expression of the transmitted laser striking through the ruby stone is as follows. Firstly, as the source of light enters the black box, I_i (the incoming light intensity) is reduced due to the reflection at the air–ruby stone interface as depicted in equation (4.3) and equation (4.4) for $n_{\text{ruby stone}} = 1.762$. I_1' is the reduced light intensity symbol after the reflection of the air–ruby stone interface. The final light intensity ratio is indicated by equation (4.5).

$$I_1' = I_i - I_{\text{reflection1}} = I_i - \left[I_i \left(\frac{n_{\text{rubystone}} - n_{\text{air}}}{n_{\text{rubystone}} + n_{\text{air}}} \right)^2 \right] \tag{4.3}$$

$$I_1' = I_i - \left[I_i \left(\frac{1.762 - 1}{1.762 + 1} \right)^2 \right] = 0.9239 I_i \tag{4.4}$$

$$\frac{I_1'}{I_i} = 0.9239 \tag{4.5}$$

Table 4.1 lists the final light intensity ratios that are based on different values of refractive index and different systems of image reconstruction (Systems A and B).

Table 4.1. Final light intensity ratio comparison for different types of ruby stone RI and image reconstruction systems.

Refractive index of ruby stone	1.762	1.770
Final light intensity ratio $\frac{I_1'}{I_i}$ (System A)	0.8478	0.8455
Final light intensity ratio $\frac{I_1'}{I_i}$ (System B)	0.9239	0.9227

4.2.2 Image reconstruction

In this study, the image is reproduced based on the LBP algorithm. This algorithm was selected for use in this research due to it being convenient and effective in processing the output image. The equation used is shown in equation (4.6) [15].

$$V_{\mathrm{LBP(160views)}}(x, y) = \sum_{tx=0}^{159} \sum_{rx=0}^{159} S_{tx,\, rx} M_{tx,\, rx} \tag{4.6}$$

In the above equation, $S_{tx,\, rx}$ refers to V_{CCD}, and $M_{tx,\, rx}$ refers to the sensitivity map [7]. Each CCD represents 40 views of the LBP image, with each view indicating the number of sensors and pixels of the CCD. For four CCDs, there are 160 views. A high number of sensor views per projection will give a higher image quality [20]. This research used the Sony ILX551A CCD sensor, which consists of 2048 pixels based on its data sheet [21]. Details on the mean value of the CCD voltage output calculation are explained by equation (4.7) for total pixels of 40 views [20].

$$V_{\mathrm{CCD,\ mean}} = \frac{\text{Total number of pixels}}{\text{Number of view}}$$

$$V_{\mathrm{CCD,\ mean}} = \frac{2048}{40}$$

$$V_{\mathrm{CCD,\ mean}} = 51.2 - 51 \tag{4.7}$$

The length per pixel for the Sony ILX551A CCD is equal to 0.014 mm [21]. Therefore, the length per pixel for 160 views is calculated in equation (4.8).

$$\text{Length per pixel}_{\mathrm{160views}} = \text{Length per pixel of CCD} \times V_{CCD,\ \mathrm{mean}}$$

$$\text{Length per pixel}_{\mathrm{160views}} = 0.014\ \text{mm} \times 51$$

$$\text{Length per pixel}_{\mathrm{160views}} = 0.714\ \text{mm} \tag{4.8}$$

4.3 Results and discussion

As described previously in the methodology section, there are two types of image reconstruction systems that shall be analyzed in this study. The first system is the optical tomography system with laser as the transmitter (System A). Conversely, the second system is the optical tomography system without laser as a transmitter (System B). For both systems, the transmitted light that strikes the CCD will be analyzed through its final measured light intensity. Finally, the simulation image output and ensuing statistical analysis will be used to validate and compare the capabilities of these two image reconstruction systems.

4.3.1 Analysis on image reconstruction

Based on the mathematical algorithm established in table 4.1, the final light intensity ratio of the ruby stone varies and is dependent on two factors: the type of image reconstruction system and the refractive index value of the ruby stone. The results for each system are presented in tables 4.2 and 4.3, respectively. A comparison between the two-dimensional images produced by each system is shown in table 4.2. From table 4.2, the two-dimensional images in (c) and (d) clearly show that a system without laser as a transmitter (System B) results in a significant difference between the higher and lower refractive index values as compared to the system with laser as a transmitter (System A) as shown in images (a) and (b).

In brief, for two-dimensional images, the difference in color intensity in the middle of the object's image in System B is clearly visualized and detected in comparison to the output from System A. However, the two-dimensional images are still a qualitative method to differentiate between rubies' clarity, and therefore further investigation on the quantitative grading of this stone is visualized and explained in three-dimensional forms presented in table 4.3.

The three-dimensional forms presented in table 4.3 show the *z*-axis analysis for the experiments with different image reconstruction systems and also differing values of the refractive index. The *z*-axis refers to the pixel values based on the multiplication results of the sensitivity maps for 160 views with the CCD normalized voltage values [15, 21]. Different levels of transparency will give a different *z*-axis, whereby a higher level of transparency will produce a lower *z*-axis, while a higher opacity level will result in a higher *z*-axis [7–10].

Table 4.2. Comparison between two-dimensional images of both systems A and B.

Refractive index value	1.762	1.770
Two-dimensional image output with laser as transmitter (System A)	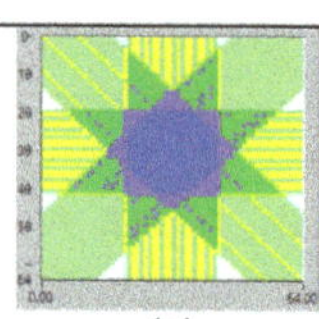 (a)	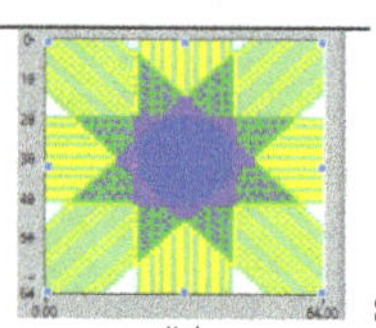 (b)
Two-dimensional image output without laser as transmitter (System B)	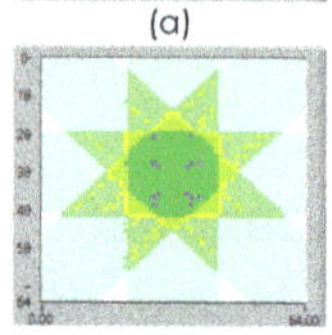 (c)	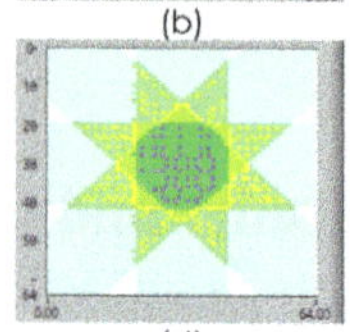 (d)

Table 4.3. Comparison between two-dimensional images of both systems A and B.

Refractive index value	1.762	1.770
Three-dimensional image output with System A	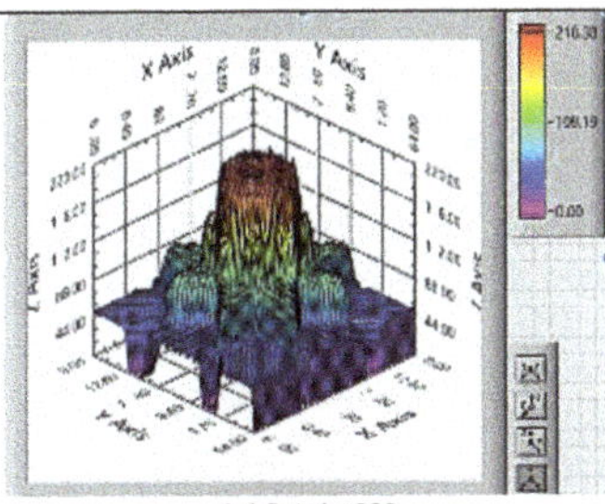 (a) Z-Axis<220	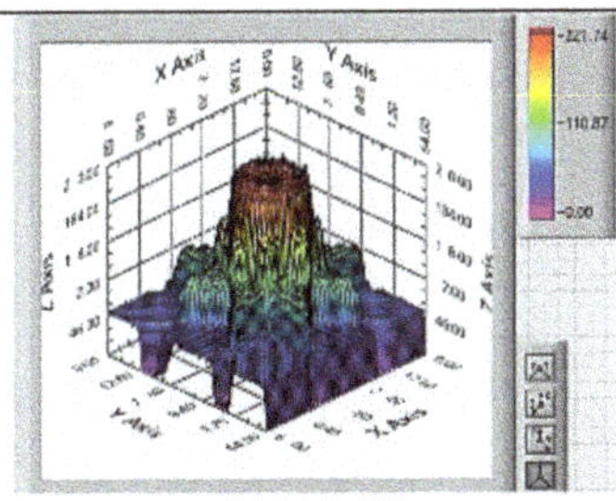 (b) Z-Axis>220
Three-dimensional image output with System B	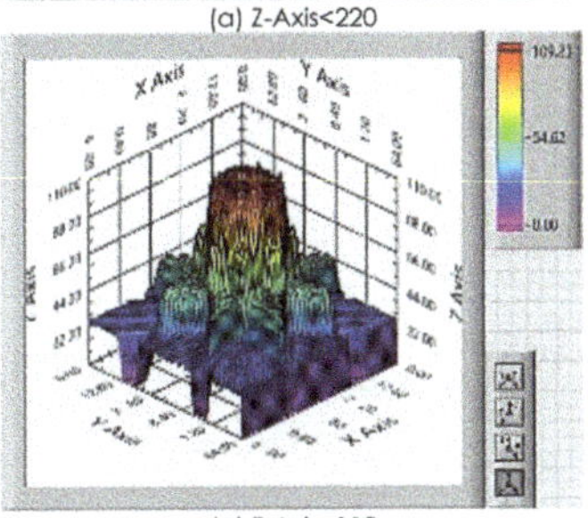 (c) Z-Axis<110	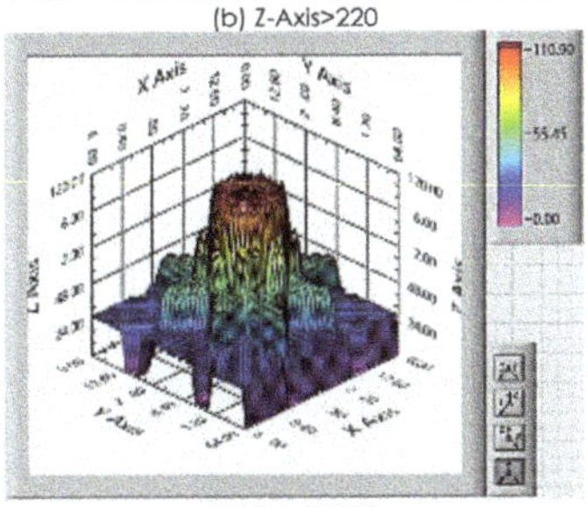 (d) Z-Axis>110

Both three-dimensional images for System A and System B are validation that a lower refractive index value of ruby stones also gives a lower value of z-axis. As a result, a significant difference in the z-axis can be observed between the maximum and minimum ranges of refractive index as given by the Gemology Tools Professional [22]. A comparison between the three-dimensional images of both systems is shown in table 4.3.

4.3.2 Statistical ANOVA test analysis

In this section, a discussion is made on the comparison between the lower and higher RI of ruby stones in two different image reconstruction systems. The analysis involves the z-axis generated from the three-dimensional images of two different values of the lower and higher RI, which are 1.762 and 1.770, respectively [22]. The main objective of this statistical analysis is to validate the potential capability of CCD simulation systems to distinguish the types of ruby stones based on their refractive index values and by using the LabVIEW programming software. The hypotheses posited in this study are as follows:

H_0: *mean value of ruby stone with lower refractive index = mean value of ruby stone with higher refractive index.*

H_1: *mean value of ruby stone with lower refractive index ≠ mean value of ruby stone with higher refractive index.*

The null hypothesis H_0 states that no differences are detected by the optical tomography system when measuring the higher and lower RI values. On the other hand, the alternative hypothesis H_1 states that differences are detectable between the two. Figures 4.3 and 4.4 depict the graph results of the ANOVA test and the data

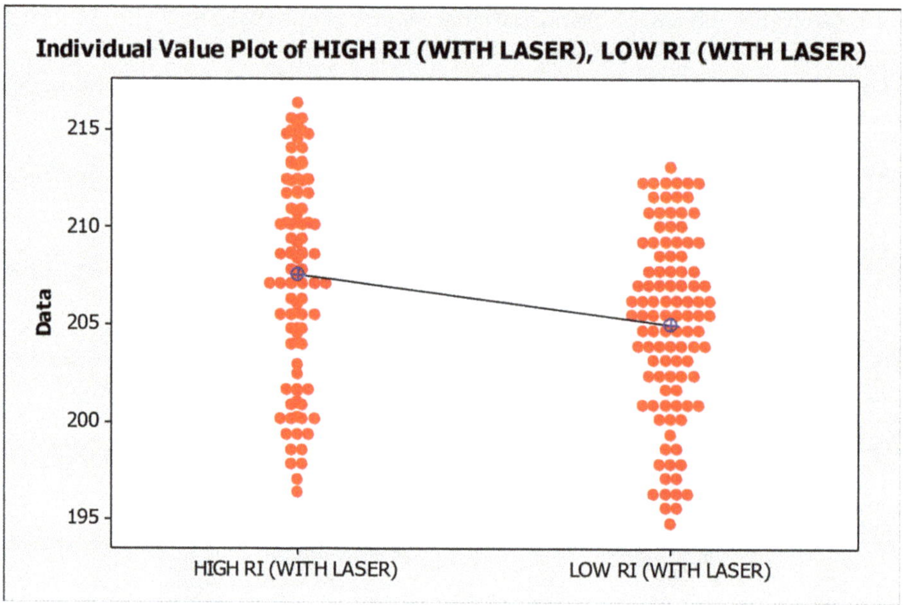

Figure 4.3. ANOVA test graph results for image reconstruction system with laser (System A).

One-way ANOVA: HIGH RI (WITH LASER), LOW RI (WITH LASER)

```
Source   DF      SS      MS      F      P
Factor    1   337.5   337.5  13.76  0.000
Error   206  5051.0    24.5
Total   207  5388.5

S = 4.952   R-Sq = 6.26%   R-Sq(adj) = 5.81%



Level                  N    Mean  StDev
HIGH RI (WITH LASER)  104  207.57   5.22
LOW RI (WITH LASER)   104  205.03   4.67

                           Individual 95% CIs For Mean Based on
                           Pooled StDev
Level                      ---------+---------+---------+---------+
HIGH RI (WITH LASER)                            (-------*-------)
LOW RI (WITH LASER)        (-------*-------)
                           ---------+---------+---------+---------+
                                 205.2     206.4     207.6     208.8

Pooled StDev = 4.95
```

Figure 4.4. Data summary of one-way ANOVA for image reconstruction system with laser (System A).

summary for System A. Figure 4.3 indicates that there is no overlap between the two RI values of ruby stones detected by the optical tomography system. This validates and further demonstrates the capability of this system to distinguish between two different RI values of ruby stones for image reconstruction systems with laser as a transmitter. Based on figure 4.4, the mean value of the higher RI is equal to 207.57, and for the lower RI, it is equal to 205.03, with a standard deviation equal to 5.22 and 4.67, respectively. At a confidence interval of 95%, the mean and standard deviation values of both groups do not overlap with each other.

More importantly, the *P*-value is determined to be equal to 0. Since it is lower than 0.05, this results in the rejection of the null hypothesis for the image reconstruction system consisting of Laser–Ruby Stone–CCD. It is hereby concluded that the optical tomography system is capable of measuring the difference between two distinct RI values in an image reconstruction system that is composed of laser as a transmitter.

Figures 4.5 and 4.6 are the graph results of the ANOVA test and also the data summary for System B. Figure 4.5 shows that there is no overlap between the two RI values of ruby stones detected by the optical tomography system. This validates and proves the capability of optical tomography systems to make the distinction between two different RI values of ruby stones for image reconstruction systems without laser as a transmitter. Based on figure 4.6, the mean values of the higher and lower RI are 103.38 and 101.98 with standard deviations equal to 2.62 and 2.21, respectively. With the confidence interval set at 95%, both groups of mean and standard deviation values do not overlap each other.

The *P*-value is equal to 0. Being lower than 0.05, this entails rejecting the null hypothesis for this image reconstruction system that is composed of Ruby Stone–CCD. It is therefore concluded that the optical tomography system is capable of measuring the difference between two different RI values even in a system that is devoid of laser as a transmitter.

In this section, the same RI value of 1.770 was selected and compared in two different image reconstructions of optical tomography systems, one with laser as a

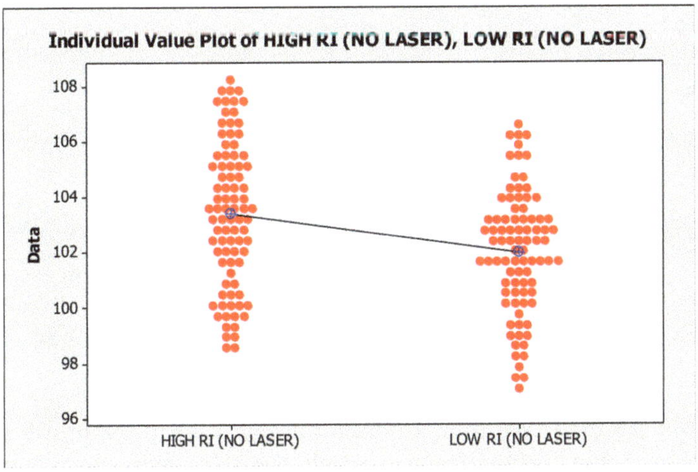

Figure 4.5. ANOVA test graph results for image reconstruction system without laser (System B).

One-way ANOVA: HIGH RI (NO LASER), LOW RI (NO LASER)

```
Source    DF       SS     MS      F      P
Factor     1    84.44  84.44  14.41  0.000
Error    170   996.26   5.86
Total    171  1080.70

S = 2.421   R-Sq = 7.81%   R-Sq(adj) = 7.27%


Level                N    Mean  StDev
HIGH RI (NO LASER)  86  103.38   2.62
LOW RI (NO LASER)   86  101.98   2.21

                    Individual 95% CIs For Mean Based on Pooled StDev
Level                  -+---------+---------+---------+--------
HIGH RI (NO LASER)                          (-------*------)
LOW RI (NO LASER)      (-------*------)
                       -+---------+---------+---------+--------
                    101.50    102.20    102.90    103.60

Pooled StDev = 2.42
```

Figure 4.6. Data summary of one-way ANOVA for image reconstruction system without laser (System B).

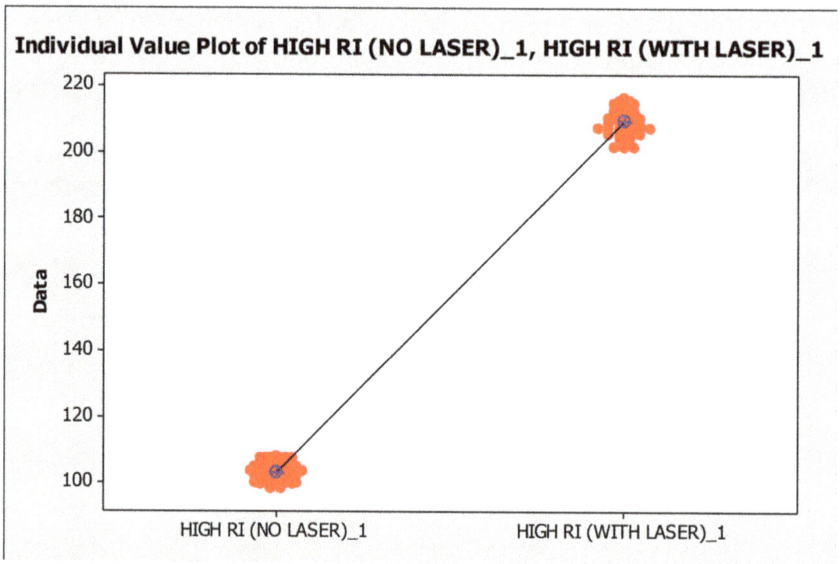

Figure 4.7. ANOVA test graph results for image reconstruction system comparison with and without laser.

transmitter and the other without. Figures 4.7 and 4.8 are the graph results of the ANOVA test and the data summary result for the comparison between image reconstruction systems with and without laser.

Individual Value Plot of HIGH RI (NO LASER)_1, HIGH RI (WITH LASER)_1

One-way ANOVA: HIGH RI (NO LASER)_1, HIGH RI (WITH LASER)_1

```
Source   DF        SS        MS         F      P
Factor    1  482373.1  482373.1  43671.53  0.000
Error   170    1877.7      11.0
Total   171  484250.8

S = 3.323   R-Sq = 99.61%   R-Sq(adj) = 99.61%


Level                   N    Mean  StDev
HIGH RI (NO LASER)_1   86  103.38   2.62
HIGH RI (WITH LASER)_1 86  209.29   3.91

                        Individual 95% CIs For Mean Based on
                        Pooled StDev
Level                   ------+---------+---------+---------+---
HIGH RI (NO LASER)_1    *)
HIGH RI (WITH LASER)_1                                      *
                        ------+---------+---------+---------+---
                            120       150       180       210

Pooled StDev = 3.32
```

Figure 4.8. Data summary of one-way ANOVA for image reconstruction system comparison with and without laser.

The hypotheses involved in this analysis are as follows:

H_0: mean of ruby stone in system with transmitter (laser) = mean of ruby stone in system without transmitter (laser).

H_1: mean of ruby stone in system with transmitter (laser) ≠ mean of ruby stone in system without transmitter (laser).

The null hypothesis H_0 proposes that there is no difference detected by the optical tomography system consisting of laser and without laser in measuring the higher RI of ruby stone. Meanwhile, the alternative hypothesis H_1 proposes that there is a difference. Figure 4.7 is the comparison of individual plots between the two image reconstruction systems, one with transmitter (laser) and the other without, and with an RI value of 1.770. Based on figure 4.8, the mean value of the high RI without laser is 103.38, while the mean value of the high RI with laser is 209.29, with standard deviations of 2.62 and 3.91, respectively. At a 95% confidence interval, both groups of mean and standard deviation values do not overlap with each other.

Here, the *P*-value is equal to 0. Being lower than 0.05, this leads to the rejection of the null hypothesis for this comparison of the said image reconstruction systems. Hence, it is concluded that there is a significant difference between the studied optical tomography systems, specifically between the system with laser as a transmitter and the other one without laser. In brief, based on this statistical analysis and the three-dimensional image in table 4.3, it is concluded that the system with laser (System A) will give an accurate image output in distinguishing the maximum and minimum RI values since the value of pixels is higher compared to System B, which results in a clearer image output.

4.4 Conclusion

In conclusion, the grading of ruby stone through its clarity characteristic has been successfully validated by the conceptual modeling proposed through the optical tomography system by using a CCD sensor and the LabVIEW programming software framework. The light from the laser penetrates the ruby stone before striking the CCD. The CCD then converts the final light intensity ratio into output voltage values. If the light intensity ratio decreases, this denotes a lower clarity of the ruby stone because of the impurities inside it. Consequently, the final light intensity received by the CCD is lesser, and the output voltage of the CCD is high. The output voltage of CCD is indicated by the color intensity of the two-dimensional image. From the ANOVA statistical testing, the image constructed with laser as a transmitter could distinguish between the different RI values of ruby stones based on the z-axis in the three-dimensional image simulation results. A ruby stone with better clarity is considerably more expensive in comparison to a ruby stone with lower clarity. In other words, a ruby stone with a lower refractive index indicates that it commands a higher price compared to a ruby stone with a higher refractive index.

Acknowledgments

The authors would like to thank the Universiti Sains Islam Malaysia and ADS research group for their cooperation in this research paper. The research is supported by the Malaysian Ministry of Higher Education Fundamental Research Grant Nos. FRGS/1/2020/WAB07/USIM/02/1 and USIM/FRGS/FKAB/KPT/53020.

References

[1] Tang S M, Tang S H, Tay T S and Retty A T 1988 Analysis of Burmese and Thai rubies by PIXE *Appl. Spectrosc.* **42** 44–8

[2] Liao K W, Lu K J, Luo R C and Yeh J A 2016 Portable dielectric tunable Forensic lens design for jadeite analysis *Int. Conf. on Optical MEMS and Nanophotonics* (Los Alamitos, CA: IEEE Computer Society)

[3] Yu J, He X and Lu Z 2019 Cause analysis of chatoyancy of sapphires from Shandong, China *RSC Adv.* **9** 24420–7

[4] Pan X, Guo Y, Liu Z, Zhang Z and Shi Y 2019 Application of cluster analysis and discriminant analysis in quality grading of jadeite red *J. Phys. Conf. Ser.* **1324** 012101

[5] Jamaludin J, Zikrillah Zawahir M, Abdul Rahim R, Mohd Yunus F R, Nor Ayob N M and Ridzuan Aw M S *et al* 2013 A review of tomography system *J. Teknol. (Sciences Eng.)* **64** 47–51

[6] Jamaludin J, Rahim R A, Rahim H A, Rahiman M H F and Rohani J M 2017 Charge-coupled device based on optical tomography system for monitoring two-phase flow tomography system for monitoring *Electron. Lett.* **53** 331–3

[7] Jamaludin J and Rahim R A 2016 Online optical tomography system for detecting and measuring the diameters of solid and transparent objects *IEEE Sens. J.* **16** 6175–83

[8] Jamaludin J, Rahim R A, Rahim H A, Fazalul Rahiman M H, Mohd Muji S Z and Rohani J M 2016 Charge-coupled device based on optical tomography system in detecting air bubbles in crystal clear water *Flow Meas. Instrum.* **50** 13–25

[9] Jamaludin J, Abdul Rahim R, Abdul Rahim H, Fazalul Rahiman H, Zarina Mohd Muji S and Shima Mohd Fadzil N *et al* 2015 Introducing an application of a charged coupled device (CCD) in an optical tomography system *J. Teknol.* **73** 3

[10] Jamaludin J, Abdul Rahim R, Abdul Rahim H, Mohd Fadzil N S, Fazalul Rahiman M H and Jumaah M F *et al* 2014 Jul 20 A review of the optical tomography system *J. Teknol.* **69** 1–6

[11] Rahalim F M, Jamaludin J, Raisin S N, Ismail I, Wahab Y A and Rahim R A 2021 Analysis on clarity of rubies gemstones using charge-coupled device (CCD).pdf *J. Tomogr. Syst. Sensors Appl* **4** 80–4

[12] Ibrahim S, Yunus M A M, Green R G and Dutton K 2012 Nov Concentration measurements of bubbles in a water column using an optical tomography system *ISA Trans.* **51** 821–6

[13] Jamaludin J, Abdul Rahim R, Rahiman M H F and Rohani J M 2020 CCD optical tomography system to detect solid contamination in crystal-clear water *IEEE Trans. Ind. Electron.* **67** 3248–56

[14] Raisin S 2020 *Labview programming for the investigation of object transparency via charge-coupled device assessment* (Nilai: Universiti Sains Islam Malaysia)

[15] Jamaludin J, Abdul Rahim R, Binti Abdul Rahim H, Fazalul Rahiman M H, Bte, Mohd Muji S Z and Mohd Rohani J 2016 Online optical tomography system application of charge-coupled device (CCD) for object detection in crystal clear water *J. Indep. Stud. Res. – Comput.* **14** 37–42

[16] Jamaludin J, Rahim R A, Rahiman M H F, Wahab Y A, Rohani J M and Sahrim M *et al* 2018 Optical tomography system using charge-coupled device for transparent object detection *Int. J. Integr. Eng.* **10** 105–8

[17] IDROAS M 2004 *A charge coupled device based optical tomographic instrumentation system for particle sizing* (Sheffield: Sheffield Hallam University)

[18] Daniel A R 1996 *An investigation into the use of dual modality tomography for the measurement of constituent volumes in multi component flows* (Sheffield: Sheffield Hallam University)

[19] Table of Refractive Indices and Double Refraction of Selected Gems–IGS https://gemsociety.org/article/table-refractive-index-double-refraction-gems/

[20] Jamaludin J, Rahim R A, Fazul Rahiman M H and Mohd Rohani J 2018 Analysis on the effect of sensor views in image reconstruction produced by optical tomography system using charge-coupled device *IEEE Trans. Image Process* **27** 1689–96

[21] ILX551A Datasheet (PDF)-Sony Corporation https://alldatasheet.com/datasheet-pdf/pdf/47512/SONY/ILX551A.html

[22] Fuller M, Smigel B W and Koivula J I 2014 PG and JWT. Gemmology Software Tools. USA

Chapter 5

Trace elements and big data application to gemology by x-ray fluorescence

Yujie Gao[1,*], Moqing Lin[1], Xu Li[2] and Xueying Sun[1,*]

[1]*Guild Gem Laboratories, Shenzhen, China*

[2]*Testing and Technology Center for Industrial Products of Shenzhen Customs District, Shenzhen, China*

The fundamentals and advantages of x-ray fluorescence (XRF) have been briefly introduced in this chapter. Gemology has evolved from the traditional stage to a higher level equipped with powerful techniques such as XRF for chemical composition analysis. We have demonstrated the broad application of XRF to gemology, including identifying the species and variety of gemstones, such as the grossular garnet tsavorite vs. the andradite garnet demantoid, as well as distinguishing the fresh pearl from the saltwater counterpart. Meanwhile, the color origin of tourmaline can be well studied with the chemical data derived from XRF. The Cr/Fe ratio calculated based on the Cr and Fe result further explains the fluorescence strength of ruby from different localities. XRF is also proven to be useful for detecting treatment, such as the dyeing treatment in pearl, which contains an abnormal amount of Ag as a strong indicator of color modification by $AgNO_3$. Unlike the traditional method, when a large amount of chemical data of gemstone from reliable sources can be tested by XRF and accumulated on a large scale, a considerable big database is built. Facilitated by the big data analysis methods, such as trace element plotting and the Linear Discriminant Analysis (LDA) method, it becomes possible to scientifically determine the geographic origin of gemstones, such as ruby and spinel, with repeatable and reliable results.

* Authors to whom correspondence should be addressed: peter.gao@guildgemlab.com, shirley.sun@guildgemlab.com.

doi:10.1088/978-0-7503-3927-8ch5

5.1 Introduction to XRF technique

X-ray is a very high-frequency, very short-wavelength and large-energy electromagnetic wave. The frequency and energy of x-rays are second only to gamma rays. The frequency range is 30 PHz–300 EHz, the corresponding wavelength is 0.01–10 nm, and the energy is 124 eV–1.24 MeV (Shackley 2011). X-ray has penetrability, but there are differences in density and thickness between different materials. When x-ray penetrates different substances, the degree of absorption is different. Different images can be obtained after the x-ray imaging processing. One typical case is the human body x-ray imaging, which has been applied in the medical field on a very large scale.

Owing to the high penetrability, x-ray can also be applied to minerals to determine their crystalline structure. Lawrence Bragg introduced the application of x-ray to a mineral and gained an image that showed the crystalline pattern, and he proposed an equation of Bragg's law of x-ray diffraction: $n\lambda = 2d\sin\theta$, where θ indicates the angle of incidence of x-rays of wavelength λ, d represents the separation of the reflecting planes, and n is an integer. This amazing work won him and his father a Nobel Prize in 1915 'for their services in the analysis of crystal structure by means of x-rays' (Thomas 2012).

5.1.1 The fundamentals of XRF

Another important property of x-ray is the fluorescence behavior, which lays the fundamentals of XRF techniques. The principles of XRF are based on the involving interactions between the electron beams and x-rays with the samples. When exposed to x-ray, the materials tested will be excited by the high energy, which will give energy to the electron within the atoms.

Gemstone materials consist of various atoms within a fixed crystalline structure. When exposed to the x-ray source, the atoms within the gemstone will be excited and emit fluorescence. Usually, the x-ray is generated in an x-ray tube, and when a primary x-ray hits an inner shell electron of an atom, it will be ejecting the electron from the atom. The empty occupation is filled by an electron from an outer shell, and fluorescence radiation is emitted, as shown in figure 5.1.

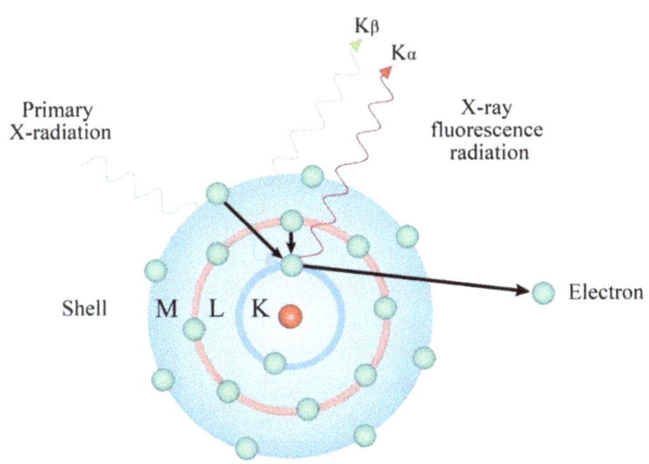

Figure 5.1. The x-ray fluorescence radiation occurs with the primary X-radiation to this the atom. The figure was illustrated by Huixin Zhao.

The conventional application of XRF first starts with geology, in which field the solid samples were the first sample type tested by x-rays. Over decades, such a technique has reached other fields, such as alloys, power samples and liquids.

XRF is currently one of the most globally accepted instrumental methods of analysis. It is widely used in environmental, industrial, pharmaceutical, forensic and scientific research areas to analyze trace chemical composition. The technique has the capability of simultaneous, multi-elemental, non-destructive analysis in a wide dynamic concentration range from 100% down to mg kg^{-1} levels with an excellent relative precision of 1% (Piorek 1994). The fluorescing atoms can be excited by energetic electrons, ions, or photons.

The fluorescent x-rays are in general measured by two types of detection systems: wavelength-dispersive XRF detection and energy-dispersive XRF (EDXRF) detection. In the former, the fluorescence x-rays are diffracted from a single crystal or multilayer optic to select a particular and often very narrow wavelength (or x-ray energy) range, often corresponding to characteristic. X-rays from an element of interest. This provides high elemental specificity and, often high sensitivity owing to the high signal-to-background ratio. EDXRF permits the detection of a broad range of elements simultaneously. The measurement resolution (elemental specificity) depends in this case on the detector x-ray energy resolution and counting rate performance (Chen *et al* 2008).

Most gemstones are consisted of minerals except for some organic material that contains certain organic substances as well. The previous study already showed that EDXRF is powerful too in the precious metal industry such as Au, Pt and Ag (Mcelroy and Mulhall 1991, Chen and Yang 1993, Yang *et al* 1995, Wang 2015). XRF is proved to be a very useful tool in gemological research and testing as the readers will see the cases in the following part. Due to the complexity of the geological setting, the chemical profiles of precious gemstones like rubies, sapphires and emeralds vary from one to another. Identification and quantification of such elements may allow us to track their localities. Similarly, the presence, absence, or abnormal amount of certain trace elements also helps us to distinguish between natural gemstones and their synthetic counterparts. EDXRF is significant in determining the authenticity of colored gemstones and their geographical origin. Compared to other equipment, EDXRF proves to be cost-efficient and perfectly suited for the non-destructive analysis of gemstones. Furthermore, it plays a significant role in determining the authenticity of colored gemstones and their geographical origin. Compared to other equipment, EDXRF proves to be cost-efficient and perfectly suited for the non-destructive analysis of gemstones (figure 5.2).

5.1.2 The advantages of XRF

The EDXRF spectroscopy is a powerful technique applied to various kinds of materials for qualitative and quantitative measurement. Several chemical analysis methods have been applied to gemology, such as Electron Probe MicroAnalysis, Laser Ablation–Inductively Coupled Plasma–Mass Spectrometry (LA-ICP-MS) and Laser-Induced Breakdown Spectroscopy (LIBS). Each method exhibits its features in efficiency and accuracy; by comparison, the advantage of XRF can be summarized as follows:

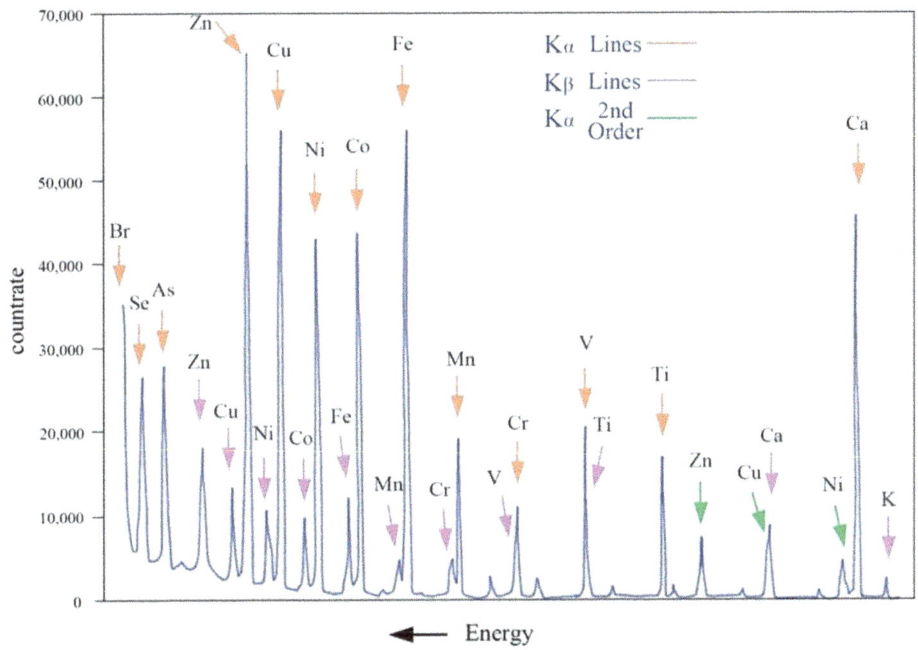

Figure 5.2. A typical wavelength dispersive XRF spectrum illustrated by Huixin Zhao.

Non-destructive: There are various methods to quantify the trace elements in the gemstones, including EDXRF, LA-ICP-MS (Liu *et al* 2013, Abduriyim *et al* 2006, Abduriyim and Kitawaki 2006) and LIBS (Krzemnicki *et al* 2004, Harmon *et al* 2009, Dubey *et al* 2021). Unlike the laser techniques LA-ICP-MS and LIBS, which evaporated a small fraction of the sample and left tiny spots on the sample, the XRF technique makes no difference to the sample and keeps the sample as it was tested before in a non-destructive way. As most gem materials are valuable and of very high price, especially for precious gemstones such as ruby, sapphire and emerald, XRF is more suitable for testing the gemstones without any weight loss in the sample.

Fast: In the XRF testing, no customized sample preparation is required; most of the gemstones can be tested and get the result in several minutes, since all the gemstones are either faceted or cabochon with a polished plane. With other methods such as Electron Microprobe Analysis, the operator has to cut and polish the sample into a flat, thin section to fit into the sample storehouse.

Clean: The testing procedure involves no chemical substance, and the x-ray irradiation is not lasting and, thus, causes no damage after the testing.

Safe: Since no environmentally hazardous chemicals are used, the XRF method is harmless. And the X-radiation shows no risk for the operator within proper protective equipment design.

Widely applied: XRF technique has been widely used in many research and industrial fields in a very broad range of applications (figures 5.3 and 5.4).

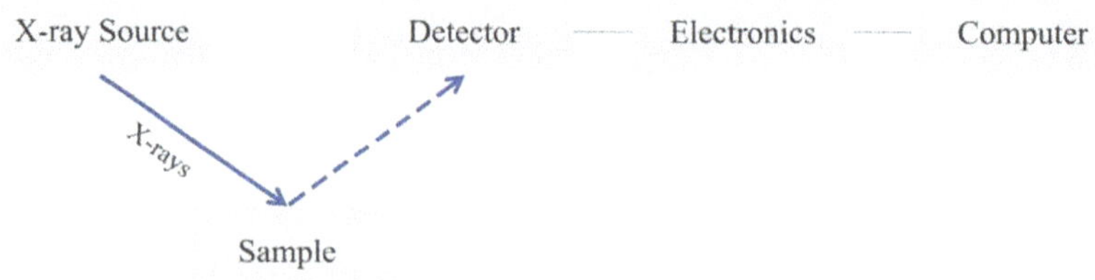

Figure 5.3. The illustration of the working flow of the EDXRF spectrometer.

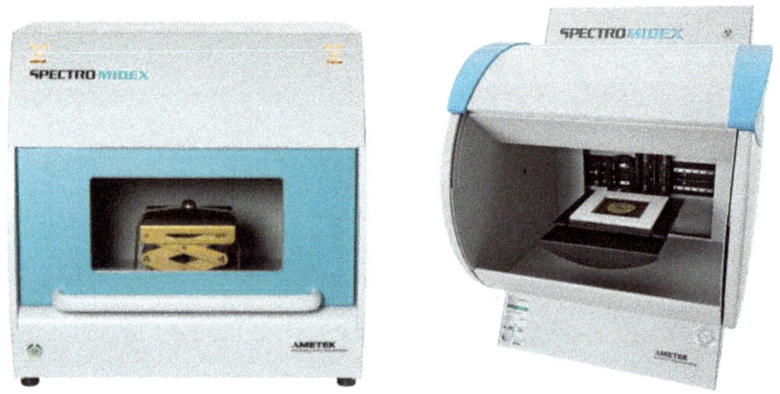

Figure 5.4. The EDXRF machine is capable of testing the gem material and providing valuable chemical results for the sample.

5.2 Trace elements and analysis in gemstones

Gemstones are valuable materials from either natural or synthetic processes, exhibiting necessary beauty, considerable rarity and decent stability. They are usually fashioned into jewelry in the gem industry. Us human beings could obtain the gem minerals from hard rocks, such as the shining Burmese ruby found in marble hosts, or from the flowing river, such as the attractive blue sapphire in Sri Lanka (Giuliani *et al* 2007, 2014, 2019). We can also culture the pearls within a proper environment, in both saltwater and freshwater. In this chapter, the term gemstone generally refers to diamond, color stones, jade and organic materials such as pearl and coral (figure 5.5).

Gemology is a scientific subject of gemstones by examining their physical and optical properties. Conventionally, the gemologist observes the gem's luster and color with their trained, naked eyes, and they can further see the internal world with a 10 × loupe or a high-powered microscope equipped with high magnification and various lighting conditions. The key refractive index of the gem can be measured by a refractometer, which also provides optical factors. Meanwhile, the density can be calculated based on the weight in the air and weight in the water of the gem using the hydrostatic weighing method.

Figure 5.5. Color stones exhibit charming and mesmerizing colors that cover the spectrum of colors. The photo was taken by Kaiyin Deng, courtesy of Guild Gem Laboratories.

However, such methods applied to gemology have been proven insufficient to meet the curiosity of the gemologist and the urged interest of the gem trade especially, as the treatment methods evolve and new gem mines and new productions have been discovered worldwide. For example, the customer not only wants to know the species of a red gem but is also interested in its geographic origin, i.e. they would like to know whether this red gem is a ruby or spinel and where it comes from. Apparently, traditional gemology seems to be helpless in the latter issue. Furthermore, the high-end techniques used to apply other fields have converged into the gemological field, which could provide gem scientists with much more information about the gems with a large amount of useful data. It is fair to say that gemology has evolved from the traditional to the modern stage of advanced level with the utilization of more techniques. What's more, facing big data, it is vital to induce the proper analysis methods for the interest of gemology at the same time. With the big data arithmetic, the information we can derive from the big data is more important than the data themselves.

In the aspect of chemical compositions, we can gain useful patterns as follows:

1. The content of one particular element can help with identification. For example, the Co content in spinel can contribute to identifying the cobalt-blue spinel from normal blue spinel.
2. The ratio between two elements by plotting, such as the Cr/Fe proves to be quite useful to explain the fluorescence behavior and facilitates the geographical origin determination to some extent.
3. The functions are generated from several sets of elements by using the LDA method. LDA is a discriminant method that attempts to model variations among samples assigned to particular groups. The method aims to maximize the ratio of the between-group variance and the within-group variance (Stanimirova *et al* 2013). The applied LDA cases in gemstones will be demonstrated in the following part.

In the following part, we will demonstrate the application of EDXRF to gemology, mainly focusing on the identification of gem species and varieties, enhancement and treatment and the geographical origin of the gemstones. Statistics methods, such as LDA based on big data of trace elements derived from the gem host by EDXRF, will also be discussed, aiming to improve the current geographic determination to a higher level.

5.3 Case study of XRF and big data gemology

5.3.1 Identification of species and varieties

5.3.1.1 Garnet group: tsavorite and demantoid

Garnet is a commonly seen mineral encountered in many rocks, and the gem-quality material is also available on a considerable scale in the gems and jewelry industry. Among the garnet family, two green species are especially favored in the gem market, i.e. tsavorite and demantoids, which both exhibit similar green color and bright luster as well as high transparency. Thus, these two gem varieties could easily get confused, and considering their difference in value in the market, it is pretty important to distinguish one from another.

The easily confused similarities and the abundant diversity of garnet minerals both result from isomorphism within the garnet group mineral. Isomorphism refers to mineral crystals composed of different chemical compositions but identical crystalline structures. For example, all the garnet group minerals consist of the same structure, which is why they are classified into the same mineral group, designating as $A_3B_2(SiO_4)_3$, where the $(SiO_4)_3$ serves as the main frame and A and B positions could be occupied by elements such as Ca, Al, Fe, etc as major elements. A wide range of isomorphism exists in a lot of minerals; for example, garnet, different species, and subspecies vary in color, chemical composition and some physical properties. A lot of gemstones, such as tsavorite, demantoid, spessartine, almandine, etc, all belong to the garnet group (figure 5.6).

Grossular is composed of Ca and Al in the A and B positions, respectively, giving a chemical formula as $Ca_3Al_2(SiO_4)_3$. When a trace amount of V and Cr replaces the Al position within the host crystalline structure, the gem host exhibits an attractive green color. By comparison, andradite consists of Ca and Fe in the A and B positions, showing a chemical formula as $Ca_3Fe_2(SiO_4)_3$. Owing to the prominent amount of Fe, andradite usually shows yellow and green colors with a darker tone. However, when V and Cr substitute Fe, the gem host will demonstrate a very nice green color with a bright tone, making it a treasure coveted in the gem market.

As demonstrated in table 5.1, the general gemological features could facilitate to tell tsavorite from demantoid. Demantoid is heavier than tsavorite due to their density difference (figure 5.7).

The refractive index value is a good indicator to quickly tell tsavorite from demantoid when using a gemological refractometer to measure their Refractive Index. We can easily get the reading value of tsavorite, while the demantoid is beyond the test range of the refractometer with a maximum value of 1.78 in most

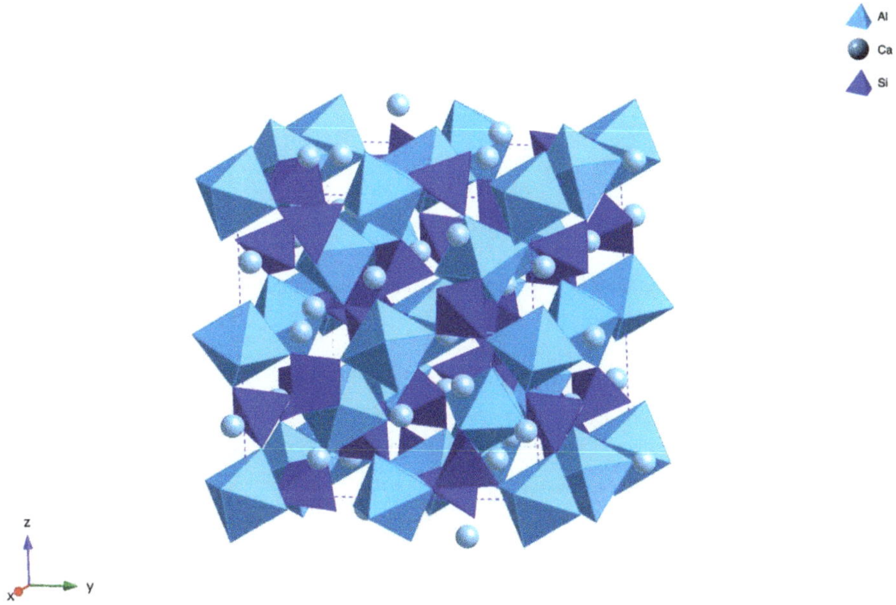

Figure 5.6. The crystalline structure of grossular (garnet).

Table 5.1. The mineralogical and gemological features of tsavorite and demantoid.

	Tsavorite	Demantoid
Mineral Group	Garnet Group	Garnet Group
Mineral Species	Grossular($Ca_3Al_2(SiO_4)_3$)	Andradite($Ca_3Fe^{3+}(SiO_4)_3$)
Color	Yellowish green to dark green	Yellowish green to dark green
Hardness	6.5–7	6.5–7
Refractive index	1.734	1.887
Optical character	Isotropic	Isotropic
Specific gravity	3.594	3.8–3.9
Occurrences	In contact and regionally metamorphosed calcareous rocks, or rocks which have undergone calcium metasomatism; in some schists and serpentinites.	In skarns from contact metamorphosed impure limestones or calcic igneous rocks; in chlorite schists and serpentinites; in alkalic igneous rocks, then typically titaniferous.
Association:	Calcite, dolomite, epidote, clinozoisite, wollastonite, scapolite, vesuvianite, diopside, tremolite, quartz.	Vesuvianite, chlorite, epidote, spinel, calcite, dolomite, magnetite.
References	Anthony *et al* (2001)	Anthony *et al* (2001)

Figure 5.7. Tsavorites and a demantoid with a similar appearance. Photos were taken by Kaiyin Deng, courtesy of Guild Gem Laboratories.

Table 5.2. Major elements of tsavorite and demantoid detected by EDXRF (ppm).

Sample	Variety	Al	Si	Ca	V	Cr	Fe
1	Tsavorite	146 100	243 100	119 900	850	151	852
2	Tsavorite	114 400	192 500	119 500	2379	316	756
3	Demantoid	6740	298 000	95 470	178	247	181 400
4	Demantoid	8400	280 200	100 900	200	257	185 700

cases. In this scenario, the EDXRF test could contribute to identifying these two species in more detailed ways.

We have selected four samples including two tsavorites and two demantoids to show the application of EDXRF to identify tsavorite and demantoids. As demonstrated in the table, the chemical composition results are listed in table 5.2, including Al, SI, Ca, V Cr and Fe. A quick comparison may find that the selected tsavorite and demantoid contain similar levels of Si and Ca, which is not surprising, since they both serve as major elements. Further analysis found the key difference lay in the Al and Fe, with tsavorite being Al-rich and Fe-poor and demantoid being the opposite, Al-poor and Fe-rich. Such prominent differences can easily differentiate and identify these two green species among the garnet group. As shown in figure 5.8, the plotting of Fe and Al could explicitly show the distinct difference of elements between tsavorite and quickly identify them by EDXRF.

The coloration agent V and Cr vary from one another, showing a fluctuating pattern due to the uneven distribution and consequently influencing the color saturation. Further study on the V and Cr could be very helpful to determine the geographical origin of demantoid as well, since the country of origin impacts the price of the gem at the same time. However, the study on the origin of garnet has not been adequate, and it is still on the research level and has not been applied in the gem industry. We will show the origin determination application to ruby and sapphire in the following parts.

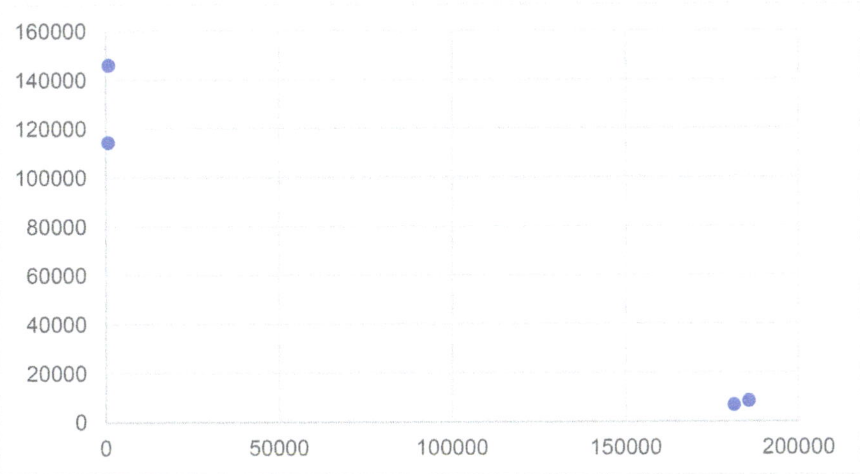

Figure 5.8. The plotting of Fe and Al could explicitly show the distinct difference of elements between tsavorite and demantoid by EDXRF.

5.3.1.2 Tourmaline: the paraiba variety

Tourmaline is a gem mineral family that is well known for producing gem-quality color stones possessing various colors that could cover the whole spectrum of color. The red color tourmaline is known as rubellite, which shows a nice color similar to the pigeon's blood color found in ruby; yellow and orange ones are coveted by the market owing to their resemblance to the canary yellow diamond. Green tourmalines showing attractive mint green and lagoon colors are also popular in the market.

The vivid neon blue to green color is also found in the Cu-bearing tourmaline, known as paraiba tourmaline, which has been quite popular for nearly three decades, ever since its discovery in Brazil (Fritsch *et al* 1990). Nowadays, such shining and neon blue treasures can be found in Brazil, Nigeria and Mozambique (Katsurada *et al* 2019). The color study on paraiba tourmaline shows that such attractive colors result from a trace amount of Cu and Mn within the gem host (Rossman *et al* 1991, Abduriyim *et al* 2006).

Owing to the prevalence of paraiba tourmaline, some 'normal' tourmaline showing similar color is sold as paraiba tourmaline, which is fraudulent and unacceptable. Hence, it is very important to differentiate paraiba tourmaline from other 'normal' tourmalines (figures 5.9 and 5.10). According to the Laboratory Manual Harmonization Committee, the very recent definition of paraiba tourmaline is as follows:

> *Paraíba' tourmaline is 'a blue (electric blue, neon blue, violet–blue), bluish green to greenish blue, green (or yellowish green) tourmaline, of medium-light to high saturation and tone (relative to this variety of tourmaline), mainly due to the presence of copper (Cu) and manganese (Mn) of whatever geographical origin* (LMHC 2012).

This definition has been widely accepted in the gem trade and at Guild Gem Laboratories. The key points of paraiba tourmaline are the color and trace element,

Figure 5.9. Paraiba tourmaline vs. blue tourmaline. The photo was taken by Kaiyin Deng, courtesy of Guild Gem Laboratories.

Figure 5.10. Blue tourmaline colored by Fe is usually mistakenly sold as Paraiba tourmaline in the trade. The photo was taken by Kaiyin Deng, courtesy of Guild Gem Laboratories.

while the geographic origin has no contribution to the nomenclature. The color can be viewed and observed by trained eyes from both seasoned dealers and experienced gem experts; however, the trace elements have to be confirmed and tested by advanced equipment. A previous study (Sun *et al* 2019) demonstrated a new method for the tourmaline species with good results. However, the LA-ICP-MS testing method is destructive and leaves a small spot on the samples; even though it is not very visible to naked eyes, it still carries a certain risk, which restricts its application to a commercial lab in daily testing (table 5.3).

Table 5.3. Chemical composition of paraiba tourmaline tested by EDXRF.

Sample	Variety	Mn	Fe	Cu	Ga	Pb	Bi
1	Paraiba tourmaline	8196	149	2430	556	533	1496
2	Paraiba tourmaline	32 780	708	19 630	12	19	318

Table 5.4. Types of pearls.

Natural/ cultured	Environment	Mollusks and pearl species
Natural	Saltwater	*Haliotis* sp. (abalone pearl) *Pinctada* sp., *Pteria* sp.
	Freshwater	*Unio* sp.
Cultured	Saltwater	*Pinctada maxima* (South Sea pearl) *Pinctada fucata* (Akoya pearl) *Pinctada margaritifera* (Tahitian pearl) *Pteria* sp.
	Freshwater	*Hyriopsis cumingii Hyriopsis schlegeli*

By contrast, EDXRF could detect the interesting element with any damage to the sample, which may be very helpful to down-side potential risk during testing.

5.3.1.3 Pearl: freshwater and saltwater

Pearls are organic gems, grown or cultivated in living mollusks. Historically, pearls (when there are only natural pearls) are quite challenging to obtain and are extremely rare. But since the early 1920s, with the development of pearl-culturing techniques, pearls are more and more available. Now, almost all pearls on the market are cultured. Both natural and cultured pearls can be found in saltwater and freshwater environments.

The chemical components of pearls include two categories: organic and inorganic components. The inorganic components are $CaCO_3$ and a small amount of $MgCO_3$, which account for about 91% of the pearl. The organic component is mainly conchiolin (a special type of scleroprotein), which accounts for about 5% of the pearl. There are also water (2%–4%) and various trace elements such as Fe, P, Cu, Co, Ag, Zn, Mg, Mo, Ni and Na. The compositions of saltwater and freshwater pearls are slightly different in trace elements and organic components (e.g. amino acids). Besides, there may be minor differences in the chemistry of pearl growth in different mollusks.

Generally speaking, freshwater (natural and cultured) pearls contain relatively higher quantities of Mn and Ba and a lower quantity of Sr than saltwater (natural and cultured) pearls (table 5.4). EDXRF can be used to separate freshwater pearls from saltwater pearls in gem laboratories. Lower Mn levels (below 20 ppm) usually indicate saltwater origin, while higher values (from 150 ppm to a few thousand)

Table 5.5. XRF chemical analysis of different types of pearls (ppm).

Pearl type	Ca	Mn	Br	Sr	Ag	I	Ba
Tahitian pearl	989 400	33.6	30.5	1182	21.8	<20	77
South Sea pearl (yellow)	989 300	83.9	42.1	1181	<20	<20	46
South Sea pearl (white)	989 300	45.6	82	1371	<20	<20	64
Akoya pearl	989 300	39.1	75.3	1140	<20	<20	59
Freshwater pearl (purple)	989 300	458.4	16	705	<20	<20	81
Freshwater pearl (white)	989 300	399.8	41	519.9	<20	<20	103

Figure 5.11. High-quality bicolor tourmalines are coveted by the gem dealers and connoisseurs in the trade. The photo was taken by Keily Deng.

indicate freshwater origin (Sturman *et al* 2019). The SrO/MnO ratio was also suggested to be used for separating freshwater and saltwater pearls, which is >12 for saltwater pearls and <12 for freshwater pearls (Karampelas *et al* 2019) (table 5.5).

5.3.2 Color origin and fluorescence

5.3.2.1 Bicolored watermelon tourmaline

An ultraviolet–visible (UV–vis) spectroscope is a valuable tool to study the colors and their relevance with trace elements. A previous study performed spectral and chemical analysis on bicolor tourmaline, as shown in figure 5.11 (Huang and Gao 2021). Two representative spectrums were illustrated in figure 5.12, with the red line representing the pink color part and the green line for the green part accordingly. The prominent band centered at 511 nm in the pink part, a tiny peak accompanied nanometer at 450 nm, both originating from Mn^{3+} (Fan *et al* 2009). The weak peak

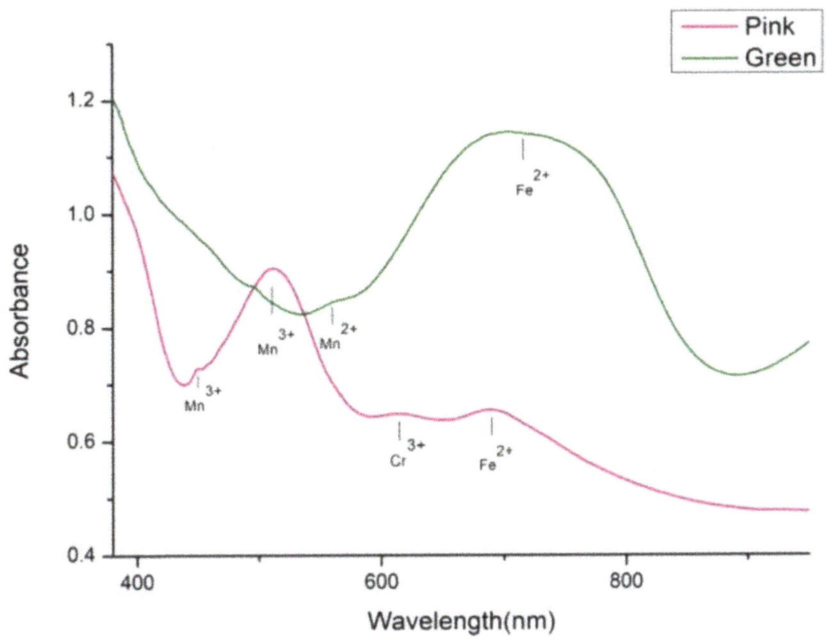

Figure 5.12. Unpolarized UV–vis spectrum of the bicolor tourmaline in this study, with the red line representing the pink color part and the green line representing the green part accordingly. The figure was illustrated by Candice Huang.

due to Cr^{3+} was positioned at 615 nm, consistent with chrome tourmaline, and the 690 nm peak was due to Fe^{2+}. Mn-related absorption takes away most of the blue and green part of the visible light, while the Fe-related peaks have little influence on the orange and red color part, thus resulting in pink color. In the green part, by comparison, the distinct broadband centered at 715 nm was caused by the Fe^{2+} absorbing most of the red color (Castaneda *et al* 2006), with only a fragile band at 560 nm attributed to Mn^{2+} (Castaneda *et al* 2006). Once again, these results prove UV–vis as powerful to unveil the mysterious origin of color in a gemstone (figure 5.12).

Tourmaline is a very complex mineral, and abundant kinds of elements are evolving in the crystalline structure. Our previous work (Huang and Gao 2021) carefully selected four elements to discuss, of which Mn and Fe are the most important. The results are presented separately based on the pink and green regions in table 5.6. No considerable Cu was detected in these samples. Cr was generally at a deficient level, insufficient to give rise to green color. Although not directly responsible for coloration, Zn is generally higher in the green part than in the pink region, with an average $Zn_{green}/Zn_{pink} = 6.74$. An interesting pattern between Mn and Fe was discovered, as shown in figure 5.13. The Fe/Mn content ratio in the pink part is designated as $Pink_{Fe/Mn}$, and green will be designated as $Green_{Fe/Mn}$. Generally, all the samples show $Pink_{Fe/Mn}$ much lower than $Green_{Fe/Mn}$, except for

Table 5.6. Chemical compositions of bicolor tourmaline by EDXRF (Huang and Gao 2021).

Sample	Pink					Green				
	Fe	Mn	Cr	Zn	Fe/Mn	Fe	Mn	Cr	Zn	Fe/Mn
1	1042	1337	35	196	0.78	31 240	17 370	44	211	1.80
2	14 480	6167	39	101	2.35	30 440	17 370	55	150	1.75
3	12 470	6152	27	96	2.03	17 510	8764	34	187	2.00
4	507	1455	15	109	0.35	25 940	11 650	41	542	2.23
5	451	1317	24	99	0.34	23 990	9638	52	771	2.49
6	425	1284	20	125	0.33	24 120	9646	39	599	2.50
7	1159	1716	19	137	0.68	21 410	8395	37	903	2.55
8	457	1392	19	68	0.33	21 380	8604	49	513	2.48
9	912	1643	20	148	0.56	18 620	6523	35	1243	2.85
10	759	1504	19	360	0.50	17 690	6585	35	1018	2.69
11	2195	1687	40	84	1.30	24 020	11 600	63	851	2.07
12	229	1187	22	20	0.19	10 880	4354	40	1131	2.50
13	454	1500	20	108	0.30	16 640	5376	73	1205	3.10
14	539	1504	19	137	0.36	18 560	6359	133	930	2.92
15	288	1179	14	78	0.24	18 370	7996	65	1274	2.30
16	396	1442	17	96	0.27	8052	3000	60	1119	2.68
17	323	1383	19	58	0.23	3986	2385	43	995	1.67
Average	2182	1991	23	119		19 579	8566	53	802	

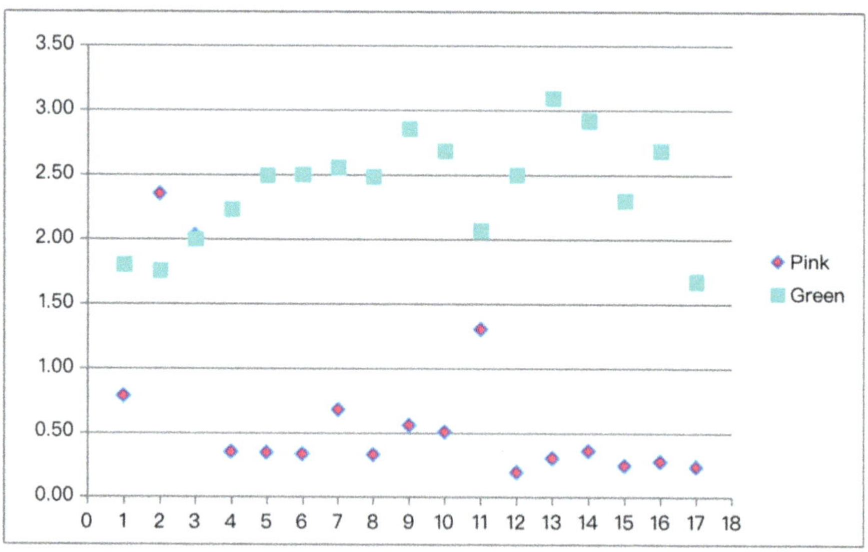

Figure 5.13. The Fe/Mn ratio of pink and green parts in 17 samples in this study (Huang and Gao 2021).

samples 2 and 3. These abnormal values found in samples 2 and 3 can be attributed to the relatively small area of pink and the large size of the testing beam of the EDXRF machine (figure 5.13).

5.3.2.2 Abalone pearl

Abalone (*Haliotis* genus) is a large marine gastropod mollusk, which can be found in the cold waters of New Zealand, Australia, Japan, China, America and South Africa. Although abalone is found all over the world, natural abalone pearls are very rare. Unlike South Sea pearls, Akoya pearls, or cultured freshwater pearls, abalone pearls cannot be cultured. The only success is a cultured blister, which is a dome-shaped bulge on the shell and sold as mabé abalone pearl in the market (figure 5.14) (Wentzell 1998). Like abalone shells, abalone pearls exhibit dynamic colors and remarkable iridescence, which are highly prized in the market. They form in different sizes and shapes, and the most common shapes are round, oval, flat baroque and horn-like baroque. Because of its high value, identification of its mollusk and color origin is important. In gem laboratories, data from different instruments are combined to reach a conclusion. One of the very useful methods is chemistry (table 5.7).

Abalone is rich in nutritious substances including iodine (Doh and Park 2018). So rich-in-iodine is a consistent property of abalone pearls. Table 5.7 is EDXRF data of

Figure 5.14. Natural abalone pearl jewelry courtesy of Mogu Gem, Baiyang.

Table 5.7. XRF chemical analysis of abalone pearl.

Element	Ca	Sr	Mn	Ba	Br	I	Ag
Content (ppm)	989 300	1530	40	bdl	60	320	bdl

bdl, below detection limit.

an abalone pearl. In addition to Ca, Sr, Mn, Br and Ba, abalone pearls also contain high content of *I* (up to 300 ppm). For other kinds of pearl, the content of iodine is usually below the detectable limit. EDXRF data does not show any heavy metal elements (such as silver), the existence of which often indicates color treatment. The high Sr and low Mn and Ba contents indicate its saltwater origin.

5.3.2.3 Fluorescence of ruby

Fluorescence is an essential gemological property of gemstones, especially in ruby. Ruby could exhibit red fluorescence with different saturations and strengths owing to the Cr content. Fe is believed to prohibit such behavior. All the rubies, both faceted and rough, showed moderate to strong red fluorescence under longwave, while they were nearly inert to weak red under shortwave. Rubies from Mozambique and Burma and synthetic counterparts by flame were observed as a comparison (figure 5.15).

When these rubies were exposed to invisible UV light, they all exhibited even red color fluorescence in longwave and shortwave UV. The strength of these rubies' reactions to longwave UV were all stronger than those to shortwave UV. The UV imaging revealed a different fluorescent appearance of these samples, which showed that the intensity of rubies of Burma origin and synthetic ones was distinctly more vigorous. In contrast, Tanzanian ruby presented a much weaker fluorescence, and Mozambique ruby appeared with the faintest fluorescence.

From EDXRF data, Fe concentration was considered to explain the different fluorescent appearances of these samples. Mozambique ruby is enriched in Fe, which will suppress the luminous intensity. Ruby from Burma and synthetic ones are characterized by a meager amount of Fe, which may induce a strong UV reaction. By comparison, Longido ruby has the highest Cr content (about 8300 ppm), and among the intermediate Fe content (about 1200 ppm), the fluorescence intensity of this origin was moderate. The Fe/Cr ratios were further proposed, presented in table 5.8, to explore the relationship between element concentration and fluorescence. It indicated that the sample with the highest Fe/Cr ratio, which was up to around 5, showed the faintest fluorescence in this experiment.

In conclusion, the different amounts of Cr and Fe and Fe/Cr ratios explain the different fluorescent appearance. It can further help discriminate rubies from various sources with different Cr and Fe content to a degree (figure 5.16).

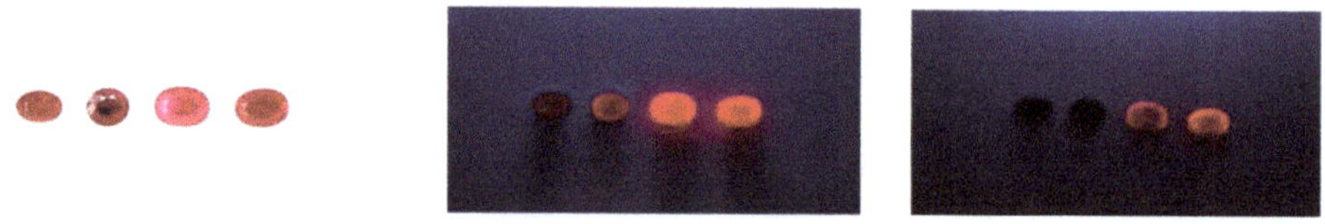

Figure 5.15. Four rubies from different origins, from left to right: Mozambique, Tanzania, Burma, Synthetic. Light source: left: daylight; middle: long wave (365 nm); right: short wave (254 nm). Photos by Xueying Sun.

Table 5.8. Comparison of Fe and Cr contents in various origins, tested by EDXRF.

Test point number	Origin	Fe	Cr	Cr/Fe
Moz-1	Mozambique	5330	1077	0.20
Moz-2	Mozambique	5246	1056	0.20
Moz-3	Mozambique	4997	974	0.19
	On average	**5191**	**1036**	**0.20**
Tan-1	Longido, Tanzania	1278	8588	6.72
Tan-2	Longido, Tanzania	1252	8361	6.68
Tan-3	Longido, Tanzania	1262	8308	6.58
	On average	**1264**	**8419**	**6.66**
Bur-1	Mogok, Burma	49	1643	33.33
Bur-2	Mogok, Burma	52	2547	49.08
Bur-3	Mogok, Burma	51	2578	50.16
	On average	**51**	**2256**	**44.35**
Syn-1	Flame Grown method	42	3233	77.72
Syn-2	Flame Grown method	46	3271	70.95
Syn-3	Flame Grown method	38	3197	85.25
	On average	**42**	**3234**	**77.48**

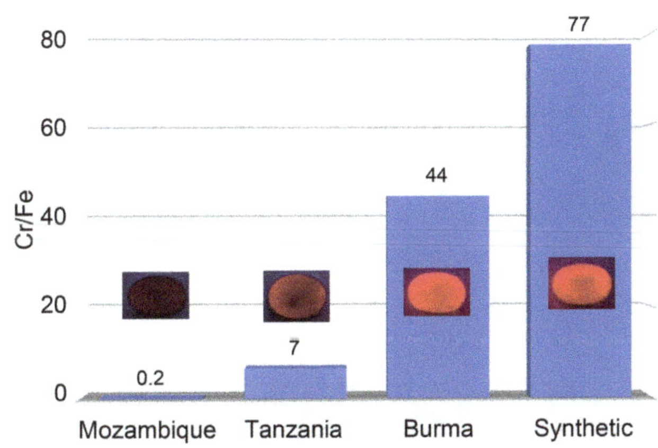

Figure 5.16. The fluorescence intensity of ruby from various origins may be attributed to the Cr/Fe ratio: the higher the Cr/Fe ratio, the stronger the fluorescence. The figure was illustrated by Yujie Gao.

5.3.3 Treatment detection

5.3.3.1 The dyeing pearl

Natural-color Tahitian pearls are among the most popular cultured pearls in the marketplace today. They come from the mollusk *Pinctada margaritifera*, and the color is from light gray to black. Treatments such as irradiation and dyeing are used to turn the less-expensive pearls black. Irradiation is hard to detect by using conventional testing instruments, while dyeing treatment usually can be identified by careful observation and spectroscopy. Some samples present identification challenges, requiring the use of chemical analysis to detect trace elements such as iodine (McClure 2010). EDXRF is useful to detect treatments with inorganic substances such as Ag, iodine, Br, etc.

Freshwater pearl might be treated with $AgNO_3$ to darken the color. The process is to immerse pearls in $AgNO_3$ solution for a while and then further expose them to strong light or put them into hydrogen sulfide gas for the reduction reaction. This process leads to the precipitation of Ag between the nacreous layers of the pearl and makes the pearls appear black.

5.3.3.2 Microscopic observation

The color of dyed pearls is generally deeper than most of the natural-color Tahitian pearls with metallic luster and a strong multi-colored overtone. The dyed pearls can also appear brownish. Figure 5.17 shows three dyed freshwater pearls. These pearls can be identified by their unusual appearance and dye concentrations in surface pits/imperfections and along drill holes (figure 5.18).

5.3.3.3 Spectroscopy

UV–vis–near-infrared (NIR) spectra are very helpful to identify dyed freshwater pearls as well. The UV–vis–NIR spectra of dyed pearls exhibit various patterns, and all are different from 'typical' Tahitian pearls (figure 5.19). Besides, 280 nm absorption of dyed pearls is much weaker.

Figure 5.17. Three dyed freshwater pearls. Photos were taken by Mengdi Yang.

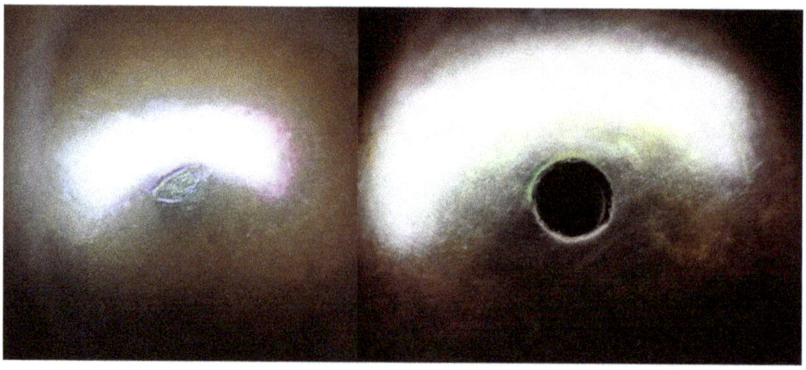

Figure 5.18. Color concentrates in the pits and along the drill hole, which indicates that pearls have been treated. Photomicrographs were taken by Huixin Zhao.

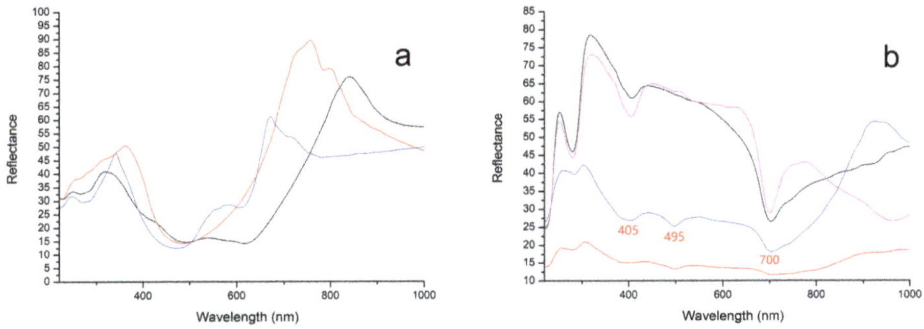

Figure 5.19. The UV–vis–NIR spectra of dyed freshwater pearls (a) and natural-color Tahitian pearl (b). The peaks at 405, 495 and 700 nm are characteristic peaks of Tahitian pearls.

Table 5.9. The EDXRF results of dyed freshwater pearls.

	Mn	Br	Sr	Ag	I	Ba
D 01	964	52.2	601.3	599.5	<20	140
D 02	799	27.7	561.6	581.7	<20	155
D 03	423.8	23	247.4	750.1	<20	105
D 04	935	37	297.4	309.6	<20	81

5.3.3.4 Chemistry

The XRF results of dyed pearls showed that the content of Ag was between 310 and 750 ppmw. The Ag ions in the $AgNO_3$ dye are first adsorbed to the surface of the pearl and then slowly penetrate the entire nacre through the tiny voids and micropores. The presence of Ag proves that they were dyed with $AgNO_3$ (table 5.9).

5.4 Big data application in gemology: geographic origin determination

5.4.1 Ruby

5.4.1.1 Introduction of ruby

Ruby, derived from the Latin word 'ruber,' means 'red.' It is the birthstone for July, and it symbolizes love, romance and passion. In ancient times, the Burmese believed it brought protection and invincibility to the one wearing it.

Ruby is the red variety of gem quality mineral 'corundum' and the sister of sapphire. Corundum is comprised of Al_2O_3 and is colorless when pure. A minute amount of Cr ions (Cr^{3+}) in its crystal lattice gives the gemstone a gorgeous red color. Ruby, sapphire and emerald are collectively called the 'big three.' Ruby, the second-hardest gemstone after diamond, bears a Moh's hardness of 9, which enables it to be durable enough for daily wearing. Ruby's beauty, durability and rarity all contribute to its status of being a precious stone and one of the most coveted red stones in history.

Rubies can be found in several different countries, including but not limited to Burma (Myanmar), Afghanistan, Vietnam, Thailand and East African countries such as Mozambique, Tanzania, Kenya, etc. The quality of a ruby is determined by its color, cut, clarity and carat weight. The most sought-after color is a pure and saturated red known in the trade as 'pigeon's blood.' Each gemological laboratory has its color-grading system and definition of pigeon's blood. The color of rubies can be altered through treatment. Heating is a commonly used and accepted treatment in the trade. The majority of the rubies are heated to gain or improve their colors. Thus, a one-carat unheated pigeon's blood ruby should be considered uncommon, a three-carat unheated pigeon's blood ruby is rare, and an unheated pigeon's blood over five carats is truly extraordinary.

5.4.1.2 The crystal structure and trace elements of ruby

Ruby, one of the most attractive precious red gemstones, is a variety of corundum. The chemical composition of corundum is Al_2O_3. The only cation (positive ion) in pure corundum is Al^{3+}, and the only anion (negative ion) is O^{2-}. It consists of Al–O octahedrons lying in planes approximately parallel to (0001). These octahedrons are linked together both laterally and vertically, forming a three-dimensional framework. Neither Al^{3+} nor O^{2-} arrayed in the corundum lattice structure absorbs light in the visible region of the spectrum. The color of corundum gems is therefore determined by trace element impurities, such as Cr, Fe, etc (Emmett *et al* 2017).

5.4.1.3 The geographic origin determination of ruby

The corundum consisting of pure Al_2O_3 is colorless. The Al position within the crystalline structure can be substituted by various elements, which absorb part of the visible light and result in red color, known in the trade as ruby. However, rubies showing very similar colors may possess highly varied trace elements due to the geological formation. Such a subtle variation could further facilitate telling the source of ruby. As shown in figures 5.20 and 5.21, the Al atoms are usually replaced by Fe, Cr, V, Ti and Ga. Advanced equipment could detect these trace elements within a ruby structure by parts per million (ppm). The term 'ppm' is a measurement unit that means dividing the ruby into 1 million parts and calculating how many parts of the specific element are in the ruby. For example, 1700 ppm Cr represents 1700 parts out of 1 million parts of ruby consisting of Cr.

Usually, the marble-hosted ruby is poor in Fe, while the amphibole-hosted and basalt-related ones are rich in Fe. Hence, the Fe content is a quick path to tell whether ruby is marble-hosted or not. Ti is found to be much higher in Mong Hsu ruby and some Vietnam ruby compared with ruby from Mogok, even though these three localities are all marble-dominated. Readers with some background knowledge of the color mechanism of blue in corundum will not be surprised. Blue zonings are found both in the Mong Hsu and some of Vietnam, and such zoning is colored by the charge transfer between trace amounts of Fe and Ti. Hence, it is not surprising to detect high Ti in blue zoning–contained rubies since Ti is one of the main coloration agents of blue color. Furthermore, although less-mentioned, V is also a contributor to the red color of a ruby. The V-rich ruby is currently found in

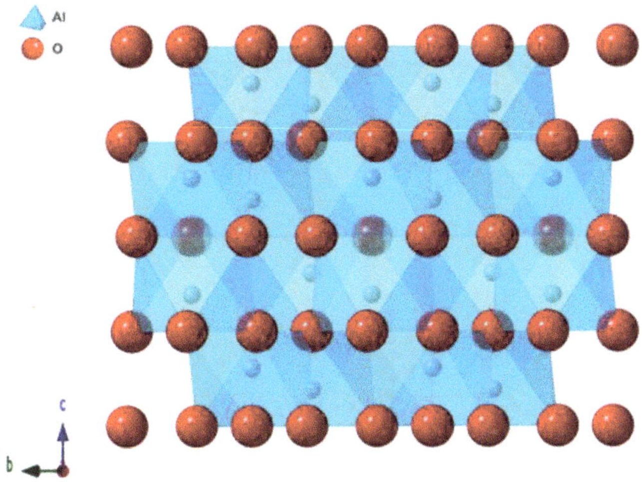

Figure 5.20. The arrangement of atoms in the corundum structure from the *a*-axis shows the Al–O octahedrons layers.

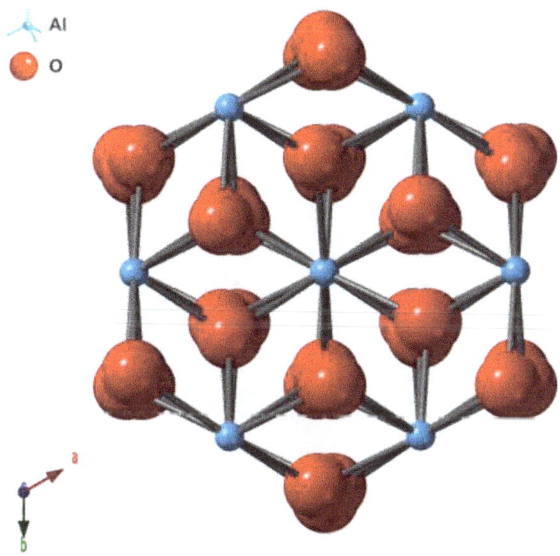

Figure 5.21. The hexagonal basal plane is shown from the *c*-axis, showing the trigonal crystal system of the corundum.

Burma only, which serves as a helpful indicator to separate Burma from other marble-hosted origins.

However, simply glancing at the massive data may get the readers lost; the analysis tool widely used in the geochemical field could clarify the numbers into a visual diagram. As shown in figure 5.22 as one example from a previous study (Gao and Sun 2021), the Fe–Cr plotting diagram displays a clear boundary that helps discriminate the ruby origins (figure 5.23 and table 5.10).

Figure 5.22. Fe–Cr chemical plotting of ruby from Mozambique, Burma and Tanzania (Gao and Sun 2021).

Figure 5.23. Top-quality Mozambique ruby. The photo was taken by Huixin Zhao, courtesy of Guild Gem Laboratories.

Table 5.10. Selected EDXRF data for the ruby samples from Burma, Tanzania and Mozambique for comparison (ppm).

Burma			
	V	**Cr**	**Fe**
Range	38–489	291–6201	39–280
Average	251	3656	74
	V	**Cr**	**Fe**
Range	32–138	1778–16 900	1280–2920
Average	90	7010	1982
Mozambique			
	V	**Cr**	**Fe**
Range	32–205	381–4666	3933–5950
Average	105	2236	4644
Detection limit			
	V	**Cr**	**Fe**
	35	10	25

bdl, below detection limit.

We have demonstrated how advanced technology helps us to determine the geographic origin of ruby in detail. The various gem properties do not exist as parameters in isolation but should be regarded as a whole organic corresponding to each other; one phenomenon or property can support the other. Systematic understanding and elaboration of numerous phenomena and properties become increasingly important with the advancement of testing technology and the discovery of new gem mines, such as the most prominent discovery of Mozambique ruby mines.

5.4.2 Spinel

5.4.2.1 Introduction of spinel

The growing popularity of ruby alternatives rekindled appreciation for spinel's rich red color and history. Until recently, spinel was underappreciated with little recognition. Spinel was recently added as an August birthstone, sharing this month with peridot and sardonyx. In ancient times, the mines of central and southeast Asia yielded exceptionally large spinel crystals. These fine stones became known as Balas rubies, and some were the treasured property of kings and emperors, often passing through many hands as spoils of war. As a result, some of the world's most illustrious 'rubies' are spinels, one of the most famous examples is the so-called 'Black Prince's ruby.'

Geologically, gem-quality spinels are recovered from various geological settings, such as Mogok (Burma), Kuh-i-Lal (Tajikistan) and Luc Yen (Vietnam) along the Himalayan Mountain belt (Malsy *et al* 2012), as well as Sri Lanka and Madagascar

Figure 5.24. Burmese spinels are obtained from secondary deposits by the river (left). Roughs are sorted afterward (right). Photos are courtesy of Xiaotao Wei.

Figure 5.25. Belonging to the cubic crystal system, spinels (left) may exhibit perfect octagonal crystal habits similar to those of diamonds (right). The photo is courtesy of Xiaotao Wei.

(Shigley *et al* 2010). A discovery of metacarbonated-hosted spinels in Baffin Island, Nunavut, Canada, has been reported recently (Belley and Groat 2019). In the mining area in Burma, the spinels are obtained from secondary deposits by the river, and the roughs are sorted afterward as shown in figures 5.24 and 5.25.

5.4.2.2 The crystal structure and trace elements of spinel

Spinel is an oxide mineral that belongs to the cubic crystal class and is usually represented by the formula AB_2O_4, where A ions are on tetrahedrally coordinated sites and B ions are on octahedrally coordinated sites (figure 5.26). Most gem spinels

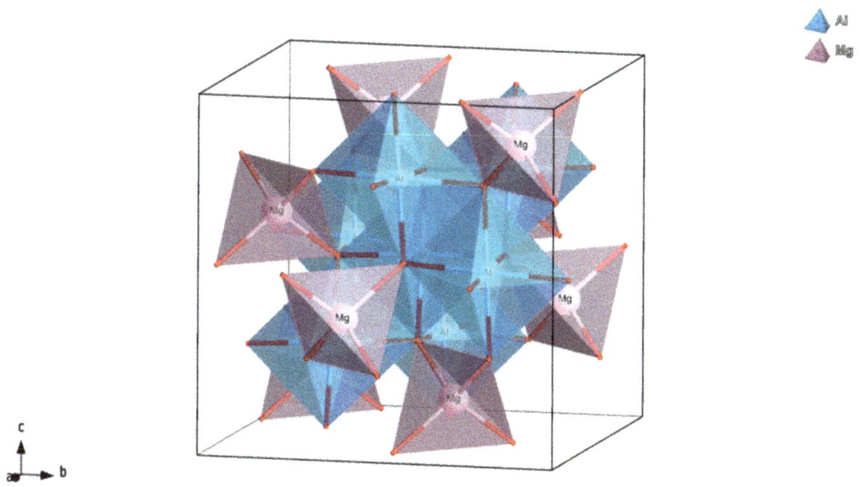

Figure 5.26. The crystalline structure of spinel.

are primarily of the typical spinel $MgAl_2O_4$ type (Saeseaw *et al* 2009). The position of Al and Mg could be substituted by other trace elements, such as Cr, Fe and Zn. Spinel is a gemstone variety rich in colors; except for yellow and green, other colors in the spectrum can appear in spinel. The rich color facilitates the use of spinel as a gem material in jewelry. Among them, the market value of pink-red spinel is the highest among all color series. Spinel bears a Moh's hardness of 8.5, which enables it to be durable enough for daily wearing. Its beauty, durability and rarity all contribute to its status of being a precious stone.

5.4.2.3 The geographic origin determination of spinel

As the ambient environments vary from one to another, the variety and content of trace elements could be a helpful indicator to distinguish the origin of spinels.

Many cations could replace Al or Mg in the crystalline structure of spinel. A previous study (Zhen and Gao 2020) demonstrated the trace elements of spinel from three localities by selecting Ti, V, Cr, Fe and Zn, and the testing results are shown in table 5.11. Generally, the samples from Vietnam were characterized as having relatively high concentrations of Fe content, ranging from 797 to 18 950 ppm. Most spinels from Tanzania seemed to have an abundance of Zn content, as high as 9700 ppm. The chromophore element Cr of Burmese spinel varied from 820 to 12 950 ppm, which is consistent with their high-saturated color (figure 5.27). A plot of Fe vs. Zn content and Cr vs. Fe can help separate these three origins, even with a small overlapping area (figure 5.28).

Preliminary statistical analysis was carried out based on the trace element's composition that contributes to the origin discrimination, such as Cr, Fe, Zn, V, etc. Subsequently, in most cases, the origins of pink-red spinels from Burma, Tanzania and Vietnam were able to be determined through the scatter plot derived from multivariate statistical analysis (figure 5.29).

Table 5.11. Selected EDXRF data (ppm) for the pink-red spinel samples from Burma, Tanzania and Vietnam (Zhen and Gao 2020).

Burma					
	Ti	**V**	**Cr**	**Fe**	**Zn**
Range	81–7083	84–1563	820–12 950	143–4640	bdl–3802
Average	917	476	4706	495	497
Median	915	446	4307	357	173
Tanzania					
	Ti	**V**	**Cr**	**Fe**	**Zn**
Range	43–302	418–1213	1013–2746	1210–4840	678–9722
Average	120	738	1810	2845	5513
Median	106	718	1699	2688	5975
Vietnam					
	Ti	**V**	**Cr**	**Fe**	**Zn**
Range	70–3365	42–1225	392–5862	797–18 950	57–4844
Average	337	352	1936	7044	1942
Median	212	324	1436	5852	1380
Detection limit					
	Ti	**V**	**Cr**	**Fe**	**Zn**
	20	35	10	25	10

bdl, below detection limit.

Figure 5.27. Pink-red spinels from Burma (left), Tanzania (center) and Vietnam (right). Photos were taken by Cuiling Zhen, courtesy of Guild Gem Laboratories (Zhen and Gao 2020).

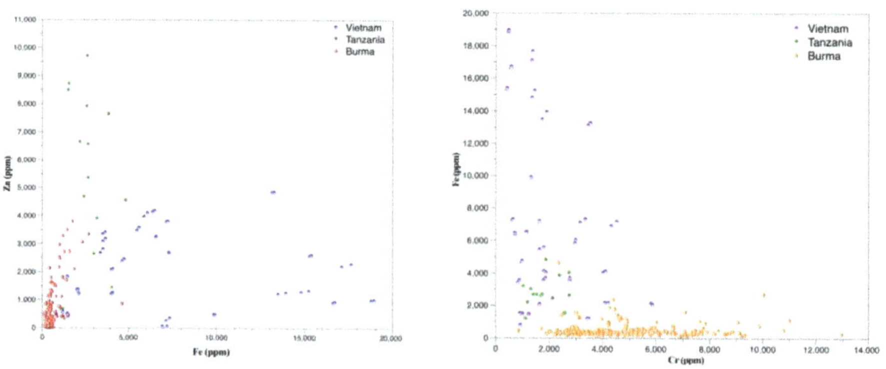

Figure 5.28. Plotting of trace elements detected from pink-red spinels from Burma, Tanzania and Vietnam: Zn vs. Fe (left) and Fe vs. Cr (right), all in ppm.

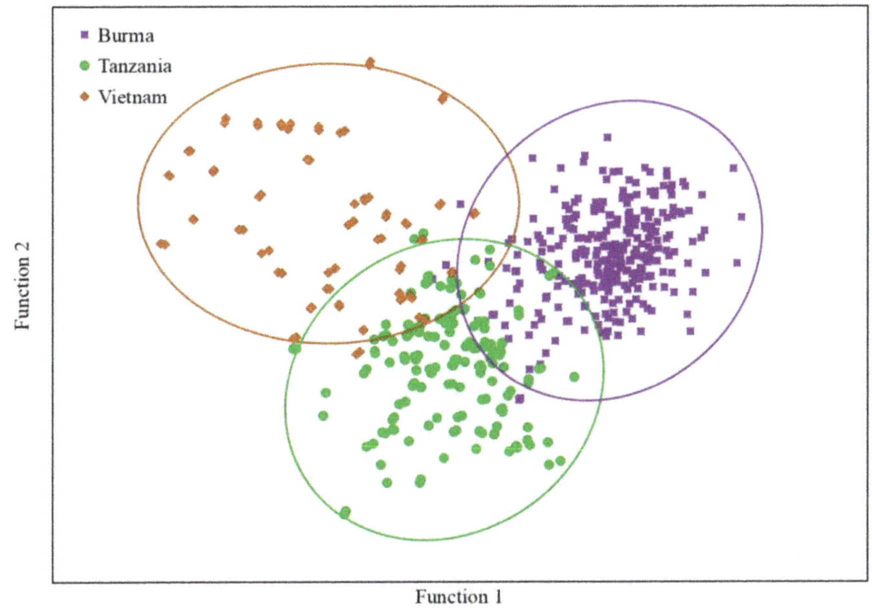

Figure 5.29. Discrimination score plot of spinels from Burma (purple), Tanzania (green) and Vietnam (orange). It can be speculated from the figure that the three origins still overlap, while about 85% of samples can be distinguished.

References and further reading

Abduriyim A, Kitawaki H, Furuya M and Schwarz D 2006 'Paraíba'-type copper-bearing tourmaline from Brazil, Nigeria, and Mozambique: chemical fingerprinting by LA-ICP-MS *Gems Gemol.* **42** 4–21

Abduriyim A and Kitawaki H 2006 Applications of laser ablation-inductively coupled plasma-mass spectrometry (LA-ICP-MS) to gemology *Gems Gemol.* **42** 98–118

Anthony J W, Bideaux R A, Bladh K W and Nichols M C (ed) 2001-5 *Handbook of Mineralogy* (Chantilly, VA: Mineralogical Society of America) http://handbookofmineralogy.org/

Belley P M and Groat L A 2019 Metacarbonate-hosted spinel on Baffin Island, Nunavut, Canada: insights into the origin of gem spinel and cobalt-blue spinel *Canad. Mineral.* **57** 147–200

Castaneda C, Eeckhout S G, Da Costa G M, Botelho N F and De Grave E 2006 Effect of heat treatment on tourmaline from Brazil *Phys. Chem. Miner.* **33** 207–16

Chen Y and Yang Z 1993 Analysis of major and minor elements in gold jewelry by XRF modified proportional factor method *Spectrosc. Spectral Anal.* **13** 91–96

Chen Z W, Gibson W M and Huang H 2008 High definition x-ray fluorescence: principles and techniques *X-Ray Opt. Instrum.* **2008** 318171

Chukanov N V, Sapozhnikov A N, Shendrik R Y, Vigasina M F and Steudel R 2020 Spectroscopic and crystal-chemical features of sodalite-group minerals from gem lazurite deposits *Minerals* **10** 1042

Doh H S and Park H J 2018 Speciation of bio-available iodine in abalone (*Haliotis discus hannai*) by high-performance liquid chromatography hyphenated with inductively coupled Plasma-Mass spectrometry using an *in vitro* method *J. Food Sci.* **83** 1579–87

Dubey S, Kumar R, Rai A K, Pati J K, Kiefer J and Rai A K 2021 Rapid analysis of chemical composition and physical properties of gemstones using LIBS and chemometric technique. *Appl. Sci.* **11** 6156

Emmett L, Stone-Sundberg J, Guan Y and Sun Z 2017 The role of silicon in the color of gem corundum *Gems Gemol.* **53** 42–7

Fan J, Feng X, Guo S and Liu X 2009 Optical absorption spectra of tourmaline crystals *J. Chin. Ceram. Soc.* **37** 523–31

Fritsch E, Shigley J E, Rossman G R, Mercer M E, Muhlmeister S M and moon M 1990 Gem-quality cuprian-elbaite tourmalines from São José da Batalha, Paraíba, Brazil *Gems Gemol.* **26** 189–205

Gao Yujie A L and Sun Xueying J D 2021 Gemological study of rubies from Longido, Tanzania *Proc. 2021 Int. Gems & Jewelry Academic Conf.* 70–5

Giuliani G, Groat L A, Marshall D, Fallick A E and Branquet Y 2019 Emerald deposits: a review and enhanced classification *Minerals* **9** 105–68

Giuliani G, Ohnenstetter D, Fallick A E, Groat L and Fagan J 2014 The geology and genesis of gem corundum deposits *The Geology of Gem Deposits* ed L A Groat 44 2nd ed (Quebec: Mineralogical Association of Canada, Short Course Series) pp 29–112

Giuliani G, Ohnenstetter D, Garnier V, Fallick A E, Rakoton- drazafy A F M and Schwarz D 2007 The geology and genesis of gem corundum deposits ed L A Groat *Geology of Gem Deposits* 37 1st ed (Yellowknife: Mineralogical Association of Canada, Short Course Series) 23–78

Harmon R S, Remus J, McMillan N J, McManus C, Collins L, Gottfried J L, DeLucia F C and Miziolek A W 2009 LIBS analysis of geomaterials: geochemical fingerprinting for the rapid analysis and discrimination of minerals. *Appl. Geochem.* **24** 1125–41

Huang T and Gao Y 2021 Spectral and chemical analysis of pink-green bicolor tourmaline *Proc. 2021 Int. Gems & Jewelry Academic Conf.* 76–81

Karampelas S, Mohamed F, Abdulla H, Almahmood F, Flamarzi L, Sangsawong S and Alalawi A 2019 Chemical characteristics of freshwater and saltwater natural and cultured pearls from different bivalves *Minerals* **9** 357

Katsurada Y, Sun Z, Christopher B and Barb D 2019 Geographic origin determination of paraiba tourmaline *Gems Gemol.* **55** 648–59

Krzemnicki M S, Henry A H and Roy A W 2004 A new method for detecting be diffusion-treated sapphires: laser-induced breakdown spectroscopy (LIBS) *Gems Gemol.* **40** 314–22

Laboratory Manual Harmonization Committee (LMHC) 2012 LMHC Information Sheet #6: 'Paraíba tourmaline' Version 7 https://static1.squarespace.com/static/5bfbb7e6cc8fed3bb9293bf3/t/5bfe92a90e2e72555d61eec3/1543410345434/LMHC+Information+Sheet_6_V7_2012.pdf

Liu Y, Hu Z, Ming L and Shan G 2013 Applications of LA-ICP-MS in the elemental analyses of geological samples *Chin. Sci. Bull.* **58** 3863–78

Malsy A K, Karampelas S, Schwarz D, Klemm L, Armbruster T and Tuan D A 2012 Orangey-red to orangey-pink gem spinels from a new deposit at Lang Chap (Tan Huong-Truc Lau), Vietnam *J. Gemmol.* **33** 19–27

McClure S F, Kane R E and Sturman N 2010 Gemstone enhancement and its detection in the 2000s *Gems Gemol.* **46** 218–40

McElroy F C and Mulhall J M 1991 Precious metal assay analysis of fresh reforming catalyst by X-ray fluorescence spectrometry *J. ASTM Int.* 105–17

Piorek S 1994 Principles and applications of man-portable X-ray fluorescence spectrometry *Trends Anal. Chem.* **13** 281–6

Rossman G R, Fritsch E and Shigley J E 1991 Origin of color in cuprian elbaite from São José da Batalha, Paraíba, Brazil *Am. Mineral.* **76** 1479–84

Saeseaw S, Wang W, Scarratt K, Emmett J L and Douthit T R 2009 Distinguishing heated spinels from unheated natural spinels and from synthetic spinels: a short review of on-going research https://www.gia.edu/doc/distinguishing-heated-spinels-from-unheated-natural-spinels.pdf

Shackley M S (ed) 2011 *X-Ray Fluorescence Spectrometry (XRF) in Geoarchaeology* (New York: Springer)

Shigley J E, Laurs B M, Janse A J A, Elen S and Dirlam D M 2010 Gem localities of the 2000S *Gems Gemol.* **46** 188–216

Stanimirova I, Daszykowski M and Walczak B 2013 Robust methods in analysis of multivariate food chemistry data *Data Handling in Science and Technology* ed F Marini 28 (Amsterdam: Elsevier) pp 315–40

Stanimirova I, Daszykowski M and Walczak B 2013 Robust methods in analysis of multivariate food chemistry data *Data Handling in Science and Technology* vol 28 ed F Marini (New York: Elsevier) ch 8 pp 315–40

Sturman N, Otter L M, Homkrajae A, Nilpetploy N, Lawanwong K, Kessrapong P, Jochum K P, Stoll B, Gotz H and Jacob D E 2019 A pearl identification challenge *Gems Gemol.* **55** 229–43

Sun Z, Palke A C, Breeding C M and Dutrow B L 2019 A new method for determining gem tourmaline species by LA-ICP- MS *Gems Gemol.* **55** 2–17

Tiantian H and Yujie G 2021 Spectral and chemical analysis of pink-green bicolor tourmaline *Proc. 2021 Int. Gems & Jewelry Academic Conf.* 76–81

Thomas J M 2012 The birth of X-ray crystallography *Nature* **491** 186–87

Wang G 2015 Discussion about precious metal jewelry testing by X-ray fluorescence spectrometry *Gold* **36** 79–83

Wentzell C Y 1998 Cultured abalone blister pearls from New Zealand *Gems Gemol.* **34** 184–200

Yang D, Zheng R and Li Y *et al* 1995 ED-XRF analysis of gold in jewelries *J. Instrumental Anal.* **5** 1–9

Zhen C and Gao Y 2020 Origin determination of pink-red spinels from Burma, Tanzania, and Vietnam *Gemol. Front.* **1** 33–45

www.ingramcontent.com/pod-product-compliance
Ingram Content Group UK Ltd.
Pitfield, Milton Keynes, MK11 3LW, UK
UKHW051851100926
8556IPUK00005B/570

9 780750 339254